The Ocean
World of
Jacques Cousteau

Invisible
Messages

The Ocean World of Jacques Cousteau

Volume 7

Invisible Messages

THE DANBURY PRESS

*The air around this **double-crested cormorant**, like the water around the fish on which it feeds, is filled with invisible messages vital to the survival of all creatures. Animals have developd a variety of senses to detect them.*

The Danbury Press
A Division of Grolier Enterprises Inc.

Publisher: Robert B. Clarke

Production Supervision: William Frampton

Published by Harry N. Abrams, Inc.

Published exclusively in Canada by
Prentice-Hall of Canada, Ltd.

Revised edition—1975

Project Director: Steven Schepp

Managing Editor: Ruth Dugan
Assistant Managing Editor: Christine Names
Senior Editors: Donald Dreves
Rick Vahan
Assistant Editors: Robert Levine
Jill Fairchild

Creative Director and Designer: Milton Charles

Art Director: Gail Ash
Art Assistant: Martina Franz
Illustrations Editor: Howard Koslow

Production Manager: Bernard Kass

Science Consultant: Richard C. Murphy

Printed in the United States of America

3456789987

LIBRARY OF CONGRESS CATALOGING
IN PUBLICATION DATA

Cousteau, Jacques Yves.
Invisible messages.

(His The ocean world of Jacques Cousteau;
v. 7)
1. Marine fauna—Behavior. 2. Animal
communication. I. Title.
[QL122.C637 1975]   591.5′9   74-23979
ISBN 0-8109-0581-7

Contents

In the animal world the central nervous system acts like a high-speed computer, perceiving and interpreting the approach of enemies, food, or a mate, and directing the animal's body to react accordingly. Vision, as we discovered in an earlier volume, *Window in the Sea,* is limited underwater, and therefore the other senses of sea creatures must act for them as SEEING WITHOUT EYES. The degree to which the other senses are developed determines the place of each animal on the ladder of survival.

The underwater world is a very busy place with messages being transmitted constantly. WAVES OF INFORMATION (Chapter I) to and from animals and from the environment itself in the form of taste, touch, and others far more alien to us are more frequent than we humans could or would ever imagine. All this, we must remember, occurs in a medium 800 times denser than air.

Until the 19th century man thought that sponges were plants and not animals, and day by day we are still learning about THE RECIPIENTS (Chapter II) of the stimuli which flow freely through the water. Sponges, we now know, can feel and react to direct contact, but there are animals in the sea with nervous systems sophisticated enough to sense electricity in the water, pick up scents, and, in fact, sense many more things than we humans can.

Because man is a terrestrial creature, he has a great deal of trouble picking up the flow of information which courses constantly through the waters. To compensate for this insensitivity, MAN'S MECHANICAL SENSORS (Chapter III) are put into use. These include echo sounders, magnetometers, and many other highly specialized scientific creations.

Communication through sounds takes place in the sea almost as much as it does on land. VOICES AND DRUMS (Chapter IV) are used to communicate the need for or discovery of food or a mate; they are heard in times of aggression; and sometimes fish and other sea animals "talk just to hear themselves talk," much in the same way we do. Grinding teeth and thumping muscles are used to make some of the sounds, and the sounds are at least as varied as those heard terrestrially.

We have always been impressed at the ability of certain breeds of dogs, such as bloodhounds, to pick up scents at great distances. In the sea the sense of smell, though useless for man, is of great assistance. Moray eels need their sense of smell to locate the octopus, a favorite meal, but the octopus can cloud this sense, as well as vision, by squirting ink between itself and the moray. The keyhole limpet can smell the presence of certain predatory sea stars at a great distance. KNOWLEDGE THROUGH SCENT (Chapter V) keeps many animals alive.

Many things about the sea are shocking to humans. The fact that ELECTRICITY AND COMMUNICATION (Chapter VI) are related is certainly one of them. Many species of fish are capable of giving off an electrical shock for a multitude of purposes; and a torpedo ray was measured as generating discharges just under 220 volts!

Sounds are pressure waves that can be heard, but there are valuable messages for fish in inaudible very low frequency pressure waves they receive through a sixth sense with an organ called the "lateral line." By SOUNDS AND PRESSURE WAVES (Chapter VII) vital information courses through the sea.

If the seas had no bottom, no surface, no suspended organisms, or no differing layers of temperature, sound would not be reflected; we would hear no USEFUL ECHOES (Chapter VIII). But sound *is* reflected and echoes are interpreted by man and dolphin alike.

We have good reason to believe that man already had a substantial language 30,000 years ago when his brain was inferior to that of today's dolphins. IS THERE ANOTHER LANGUAGE? (Chapter IX) as we define "language," and are the pops, clicks, and whistles that dolphins make equal to "words"? Many whales "sing," and repeat the same "songs" over and over. Man is getting closer and closer to the answer, and with research will perhaps someday be able to better understand these animals through more systematic methods than intelligent guessing and observation.

Beneath the surface of the sea, there alone to experience its strange beauty and stillness, in a world we are as yet unequipped for, one quickly learns that only when VIBRATING WITH THE SEA can we be all in tune with the world.

Introduction: Seeing Without Eyes

Soon after the laws of gravity were formulated, astronomers painstakingly calculated orbits and trajectories of celestial bodies. They had time to spare and no deadlines. This is no longer the case for astronauts who are launched on rockets for interplanetary voyages. Incredibly complicated calculations must be made at once; when a capsule makes a soft landing on a foreign planet, its rockets have to be fired within milliseconds, the time lapse being a function of various data (speed, acceleration, gravity, distance, etc.), which change very rapidly. No human brain could match the requirements, and high-speed computers had to be produced and programmed for such vital reckonings.

In the animal world, on land and sea, it is the central nervous system that acts like a high-speed computer, receiving and interpreting the approach of enemies or of food or of a mate—and directing the musculature to respond appropriately. The switchboard of the system, the brain, is only as efficient as the senses that inform it. And the priority of senses is not the same on land and in the sea. In the sea the paramount land sense, sight, is rarely the most important. As we discussed in an earlier book, 100 feet is about the limit sight extends even in the clearest water. A marine animal relying mainly upon sight is imprisoned in a bubble of perception only 200 feet across. If the animal is of any size, such as a human skindiver, it will hardly ever know where it is, in what direction to point itself, or when to protect itself and from what. Unless it stays in the water but a limited time, like a diving bird, or is provided with exceptional armament, like the porcupine, a sight-dependent animal in the sea soon blunders into disaster. Marine animals have especially well developed nonvisual senses, which are able to receive "invisible messages."

Many of the "invisible messages" produced by undersea creatures cannot be read or understood by man. Many years ago we took *Calypso* into the equatorial area of the Indian Ocean three different times for the entire month of April. The two first years, for many days from dawn to dusk, the ship was escorted by dozens, sometimes hundreds, of sperm whales, spouting, breaching, and raising their great flukes in the air to sound. Herds of dolphins, nothing apparently on their minds but the joy of exercising their command of the liquid world, rode our bow waves and leapt out of the water in show-off acrobatics. The third year, in exactly the same region of the sea, all was a desert before us. We saw a few dolphins, but they nervously fled from us as though we harbored a mortal disease. What could have happened? It took us two weeks to find out. Then, miles away from the area, we encountered several groups of killer whales. We humans had no knowledge that formidable carnivores were closing in on the playground of the previous springs. But the whales and dolphins apparently knew it and cleared out.

Sound is the key information medium of the great sea mammals. The sound-producing and sensing organs of the whales and dolphins perform a function as vital for them as that of eyes for the eagles. By varying the frequencies of the vibrations they emit and correctly interpreting the reverberated sound waves as they bounce back, the big animals gain an extremely detailed picture of their surroundings in any condition of murk or darkness. With zero visibility they can judge distances, distinguish between the sizes (and probably the species) of fish in the neighborhood, find their way through the mazes and obstacle courses of jagged defiles in the submerged mountain ranges.

In spite of the acuity of the echolocation system in cetaceans, and maybe because of their near-total dependence upon it, specific seashores are periodically the stage for a tragedy of marine life: the stranding of whales. How can it be that individuals or even large groups of these brainy animals, whose navigational expertise has been demonstrated in hundreds of observations and tests, suddenly lose all sense of orientation and swim into shallows from which they cannot extricate themselves? It seems incredibly perverse when stranded whales that have pulled free by helpful seamen often swim stubbornly right back onto the beach. If we reject the hypothesis that the strandings are mass suicides, the answer may lie in a form of "auditory illusion," which can afflict whales just as optical illusions may confuse land animals. Disoriented and panic stricken, the whales may thrash around aimlessly—and finally, helpless and exhausted, die. The computer engineers say: "Any computer is only as good as the data fed into its program." No matter how clever and speedy our brains, our survival depends in the first instance on the truthfulness of our senses.

Sound is not the only important sense underwater. Fish have evolved an organ which has no parallel on land: the lateral line—sensors that pick up pressure disturbances in the waters around the animals, even if they originated very far away. Smell, touch, taste, and special senses to detect gravity as well as magnetic or electrical fields play their part. A majority of radiations remain unnoticed—even if they have an influence on our behavior. In March or April the thermotropic cells of a crocus react to signals that the world is about to get warmer ... and that the little flower should show its head: the "message" has got through that spring is near.

Man is only beginning to comprehend the range of "invisible messages" the universe produces, much less read them all. Radiation and vibrations permeate all living things; each creature has its own limited scope of perception, and man is extending his own senses with the help of instruments. The sea is just different enough from land to be loaded with helpful hints. Replacement senses are inspired by the sea. For example, British scientists have built acoustic goggles for the blind, transmitting and receiving ultrasonic signals, very much in the way dolphins do. They enable a sightless man to "see with his ears." But essentially, as man develops his capability of tuning in more and more to the myriad "invisible messages" in our universe, he expands his intellectual horizons, his sources of inspiration, and his artistic and philosophical creativity.

Jacques-Yves Cousteau

Chapter I. Waves of Information

Every undersea animal receives a constant flow of information from its environment and from other animals. These messages may indicate danger or desire for mating; they may demarcate territories or announce the presence of food. Information is transmitted by a variety of means, including light, sound, pressure waves, chemicals, touching, and electricity. Even tidal rhythms, gravity, and the characteristics of water itself impart knowledge.

To be received, light messages must be seen. But as we learned in an earlier volume in this series, *Window in the Sea,* light and vision are limited underwater.

Sound is a vibration traveling through some medium—air, water, oil, stone, or whatever. It is produced by energy that causes vibrations and movement of the molecules in a medium. These moving molecules push their neighboring molecules which in turn push their neighbors and so on. This chain reaction continues until the energy that produced the sound is absorbed and nothing is left but silence. Sound is perceived by the ear in man and by equivalent organs in other animals.

> "Information is transmitted by a variety of means: pressure waves, light, sound, touching, chemicals, and electricity."

The senses of taste and smell, although seperate, rely on the same basic principle—the reception of various chemicals. Chemicals dissolved or floating free in the ocean can be perceived by the sense of smell. The sense of taste requires that an object be physically contacted. By taste a predator knows whether or not its eyes or nostrils were correct in directing it to a definite object.

The sense of smell is extremely acute in some fish, and they employ it to learn about the water's chemistry in their area, to communicate with others, or to locate a meal. The solvent properties of water, which allow molecules to be dissolved, dispersed, and transmitted easily, have led many marine creatures to become very dependent upon their ability to smell.

The sense of touch involves physical contact between an animal and an object or substance. A great number of fish and invertebrates have long sensitive extensions of their bodies (tentacles, feelers, whiskers, or barbels), which enable them to make contact without getting too close.

There are a number of remarkable fish that are able to generate electricity and to detect electrical fields. Electrical messages can warn them of danger, advise them of approaching prey, and communicate reproductive or territorial information. Such information may also be used to detect ocean currents or objects within the water.

Almost every living plant or animal is able to sense gravity. Geotaxis is the way an animal responds to the force of gravity—how it orients in space.

Whatever the means, messages are constantly permeating the sea, and animals able to receive and use them gain great advantage over those that do not have this capability.

Ulysses and diver. Attracted by human activity, a friendly grouper, which Calypso divers named Ulysses, usually stayed near the men. Occasionally, however, the fish swam away quickly as if it had received an urgent call the divers had not perceived.

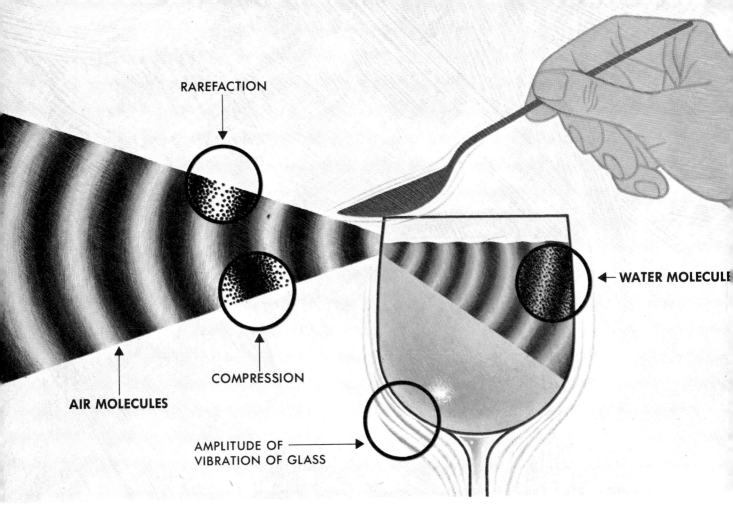

RAREFACTION

COMPRESSION

AIR MOLECULES

AMPLITUDE OF
VIBRATION OF GLASS

WATER MOLECULE

Radiating Waves

If a stone is thrown into the water, ripples are formed and travel outward from the point of impact. The concentric circles on the surface are not the only effects of the disturbance. Waves of a different sort form underwater and in the air. Those that travel through the air come to us as the sound of a stone splashing into the water. The underwater splash is felt and heard by organisms in the three-dimensional vicinity. The stone has displaced water molecules which in turn push neighboring molecules. Eventually the stone's kinetic energy is taken up by the surrounding environment, and the pressure waves become smaller and smaller until they are gone or absorbed by the bottom or shore.

Another illustration of the same effect can be demonstrated by striking the side of a glass of water with a spoon. The vibrations

Sound waves in air and water. Molecules are pushed together, causing compression. Particles behind the compression are spread in rarefaction. Distance of movement of glass is called amplitude.

of the glass also set water into motion, which produces ripples on the water's surface. The underwater waves travel faster than those in air because the molecules of water have properties relating to compressibility and elasticity that are considerably different from those of air. Although the speed of sound increases as sound travels from air to water to glass, this is not strictly a result of the increase of density.

Waves of compressed molecules—whether in air, water, glass, metal, or any other material—carry with them increased pressure. The number of them that passes a given point in a set amount of time is the frequency of the sound wave.

13

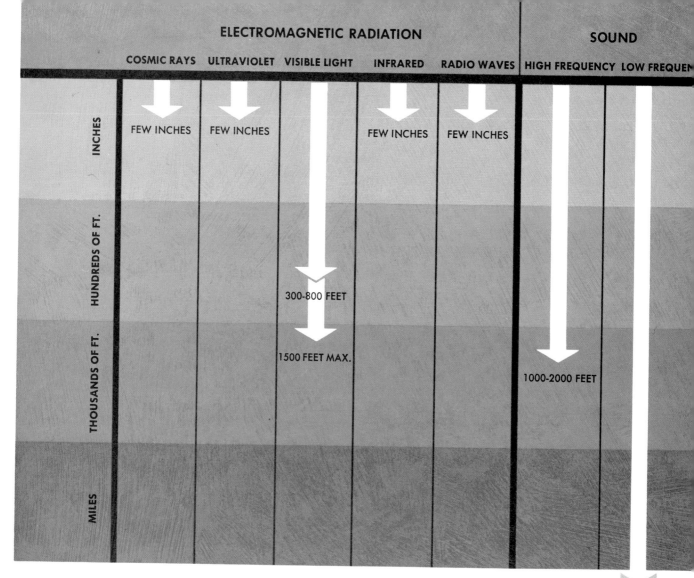

INCHES

HUNDREDS OF FT.

THOUSANDS OF FT.

MILES

FEW INCHES   FEW INCHES   FEW INCHES   FEW INCHES

300-800 FEET

1500 FEET MAX.

1000-2000 FEET

THE INTENSITY OF SOUND GREATLY AFFECTS ITS PENETRATION

MANY MILES DEPENDING ON DENSITY

Penetration of Light and Sound in Water

As light passes through water, it is scattered and absorbed. (This was discussed at length in the volume *Window in the Sea.*) As light is scattered and absorbed, it loses more and more of its components, as certain energy wavelengths do not penetrate as far as others. Thus, the shortest rays, the invisible cosmic rays, are absorbed very quickly as they pass through water. Ultraviolet and infrared rays, both of which are also invisible to the unaided eye, are also absorbed. The chart above shows that these invisible forms of light—cosmic, ultraviolet, infrared rays, and radio waves—penetrate only a few inches into water before they are absorbed. Visible light, on the other hand, penetrates to greater depths. Just how deep depends on the length of the light rays. Visible red penetrates far less than blue, a shorter wavelength. Some visible light penetrates as deep as 1500 feet in very clear water. Beneath that, all is blackness.

The penetration of sound through water similarly varies with the wavelength. As the chart indicates, the higher the frequency of the sound waves, the less they penetrate.

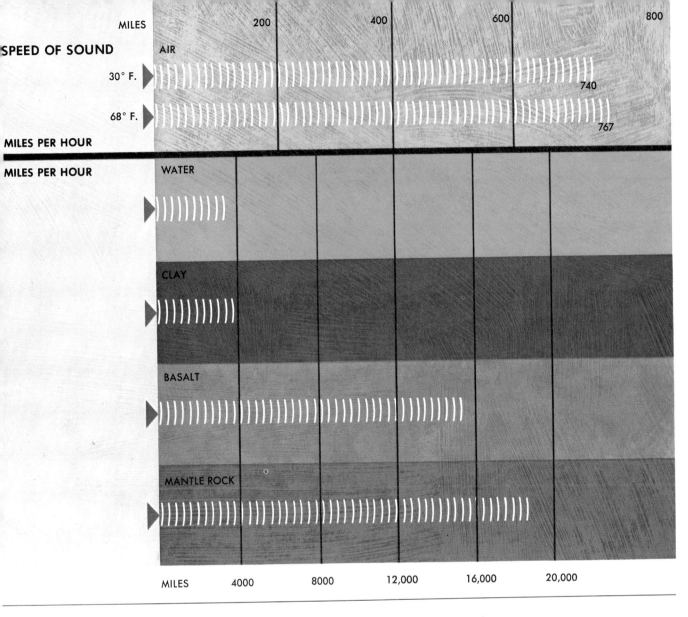

MILES	200	400	600	800

SPEED OF SOUND — AIR

30° F. — 740

68° F. — 767

MILES PER HOUR

MILES PER HOUR

WATER

CLAY

BASALT

MANTLE ROCK

MILES	4000	8000	12,000	16,000	20,000

Speed of Sound in Various Media

Sound travels through water as a series of compressions and rarefactions of the water molecules. The pop of a pistol shrimp imparts energy and causes the water molecules to bump into each other in a sequential manner. The way in which the molecules move from their original position and then return largely determines the speed of sound through that medium; that speed generally increases with the medium's density.

If the sound passes through more than one medium, it varies in speed and may be reflected, refracted, or absorbed in very much the same way as light. The speed of sound also varies with the temperature of the medium it is passing through. Thus, when air is 30° F., sound travels at the rate of 740 miles an hour, but when air is 68° F., it moves at 767 miles an hour.

In water the molecules are more confined and are to some extent bound to each other. This allows acoustic waves to pass through it at a speed of about 3500 miles per hour. Some denser substances such as basalt further enchance speed of sound, allowing it to travel over 14,000 miles per hour.

15

Diffusion

Just as sound and light are absorbed or scatter in water, so are particles of matter dispersed in the sea. The matter, whether liquid or solid, is made up of molecules, which are bound together chemically. When these molecules enter the water, they may separate from each other because of the chemical action of the water, a powerful solvent. The molecules are carried on currents or simply diffused through the water.

They can also be moved through water by electric currents. Animals in the area can smell these molecules and may be able to identify the source of the molecules from their smell. When a fish attacks another animal, the victim's blood and other fluids move with the currents and even countercurrents. Other fish in the area may pick up the scent when the individual molecules or clusters of them come to them and stimulate their olfactory sense. The fish follow the scent corridor until they track it to its source.

The same situation occurs when a diver hacks apart a sea urchin. Liquid and solid parts of the sea urchin are set adrift, and soon the whole area is teeming with small reef fish feasting on the remains of the sea urchin. In a similar manner sharks get the message of a potential meal.

Physical oceanographers use this dispersion of molecules of matter in water to study various aspects of the sea. They usually use dye markers to study tidal rips and other kinds of currents by utilizing the dis-persion principle. By tracking the dye, the scientists know where the currents go, the amount of dispersion due to eddies, and even their velocity.

Tracing water systems. Fluorescein dye released in a deep freshwater well on the island of Andros (Bahamas) finally shows green just offshore (left), proving the existence of underground channels.

Tracing water movements. A bag of rhodamine dye is released from a diver below and disperses as it rises (right). The red dye can be used to trace the movements of water in detail.

Chapter II. The Recipients

Some basic awareness is found in all creatures, but the complexity of the information an organism receives depends largely upon the complexity of the organism. In the simplest animals, individual cells respond to many stimuli, while specialized sensors are found in higher forms of life. In sponges, for instance, normal, nonspecific cells serve as sensors. They will react, as will any living

"Those senses best able to gather information develop, and those less useful disappear, degenerate, or never develop."

organism, if they are bombarded with sufficient stimuli—whether it is sound, electricity, or pressure—but the amount and quality of the information they are capable of receiving is extremely limited. For more sophisticated animals only a subtle stimulation is needed to provoke a reaction. The advantages of sensitive detection equipment are self-evident.

Competition for survival forces animals to make use of as many sources of information as they can. Some cells or tissues that possessed a low threshold to a certain stimulus had the potential of becoming a sensory receptor. If information obtained by those cells increased, the survival capabilities of the organism, specialized receptors, eventually developed.

All animals, no matter how primitive, have some means to gather the information they must have to survive. Senses develop in response to environmental factors, and since environments vary (from frigid polar waters to warm tropical seas, or from sunlit surface waters to black abyssal depths), the dominant senses of sea animals vary. Those senses that are best able to gather vital information develop most fully, and those senses that are less useful degenerate, disappear, or never develop at all. For example, animals in unusual environments (among them borers, burrowers, and animals living in turbid waters) develop special senses such as smell, touch, or electrical organs to provide the information they need to survive. These senses, therefore, have become the dominant ones for such animals.

Using their senses, all animals seek a comfortable environment, one that will not upset their internal systems. Sessile or slow-moving animals are concerned only with their immediate surroundings, so the short-range senses (touch and taste) are most important to them. However, the primary senses in faster-moving animals are long-range ones like sight, smell, hearing, and the sixth, or lateral line, sense in fish.

Whatever the senses, they help the animal decide how, when, and where to move to get food, find a mate, or keep from becoming food for another animal. In general, vertebrates depend most heavily on the long-range senses, while short-range senses are most important to invertebrates (with the exception of some cephalopods). Most animals do not rely on a single sense to guide their actions, but use the information gained from all their senses. Since various animals evaluate the information they receive differently, a great variety of life-styles has developed.

Jellyfish. A tiny chamber at the base of the tentacles of some jellyfish contains a statolith. These are the animal's balancing organs.

Sponges and Flatworms

Sponges are a collection of loosely organized, unspecialized cells. Each responds independently to stimuli. If a sponge is pricked with a sharp object, it responds by contracting, but this response is slow and may be localized. Cells more than one inch from the point of stimulation may show no reaction at all. However, if the stimulus is more general (polluted water, for example), all the sponge's surface cells may respond by closing off their pores to keep out the irritant.

Flatworm. This primitive animal (above) has an organ analogous to eyes, which is made up of tiny pigment spots. But the worm's ability to "see" is limited to the simple reception of light.

Sponges. The blue basket sponges seen in the foreground (left) are relatively insensitive creatures that have a basic cellular system which receives stimuli and reacts, albeit slowly, in turn.

Because the sponge is a simple animal, its sensory abilities are simple. It responds to the touch of a sharp object and the irritation of unpleasant chemicals, but we know little about how it perceives these stimuli.

Flatworms have primitive eyespots containing pigment, which enable them to distinguish between light and dark and, because of two receptors, determine the direction of the light. They also have two other types of sensors. One is sensitive to chemical stimuli, perceiving far-off substances by smell and sensing them on contact by taste. Another is stimulated by the passage of water over the flatworm's body surface. This is perceived by special rheotactic sensors.

The flatworm's brain is merely a swollen mass of nerves connecting two nerve cords, but it can effectively interpret information and even has the capacity to learn.

Anemone. When this animal senses the presence of a meal, its mouth at the center of the white oral disc responds. Each tentacle also responds, gathering the food and bringing it to the mouth.

Sea Anemones

Like all animals, this sea anemone is sensitive in some degree to chemical changes in its environment. When juices of food are placed close to a sea anemone, it reacts by expanding its body and waving its tentacles. When the food is removed, the reaction stops. And the sea anemone can apparently discriminate among the foods placed near it. It reacts most to meats and other forms of protein. Starches and sugars evoke only a slight response. There is virtually no response to inert objects. When food is sensed, activity begins; the tentacles reach out, grasp the food and move it to the mouth.

Gorgonian polyps. The feeding habits of gorgonian polyps indicate that they are sensitive to their environment. They react to light, the presence of food and other chemicals, and physical contact.

Gorgonians

Sensory cells are found in the inner and outer skin of the alcyonarian, or soft, corals, including the gorgonians. These corals do not build reefs out of the individual coral animals' skeletons. Sensory cells of these colonial animals project through the skin at one end and reach inward until they branch into fine fibers connected with other nerves and nerve junctions. The gorgonian's sensory cells fall into several categories: some are sensitive to light; some react to touch; and some are sensitive to chemical stimulation, reacting to substances that drift through their waters.

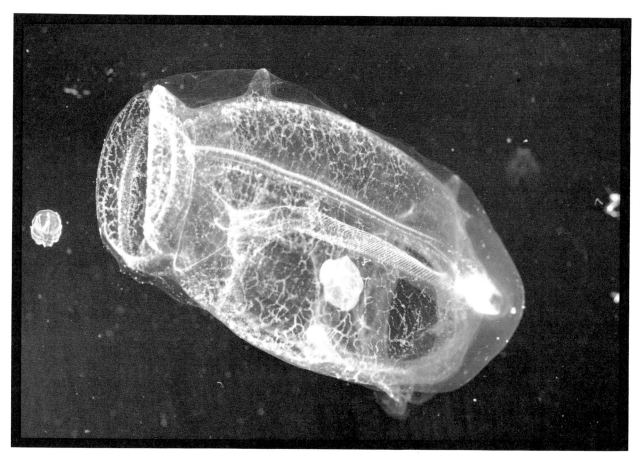

Salps

Barrel-shaped salps are classified in the same broad group of animals that men are members of—phylum Chordata. Yet they are relatively primitive creatures.

In their adulthood, salps have no backbone, brain, or spinal cord, and their sensory system is minimal. But in their larval stages they have a spinal cord as well as a concentration of nerve cells that could pass for a brain. In the course of their development into the adult stage, the salp's brain and spinal cord degenerates and disappears. What little sensory equipment they have consists of a number of tactile and chemoreceptor cells on the inner and outer surfaces of their siphons. These siphons are organs that carry water in and out of the salps for respiration and nourishment. With its tactile cells the salp recognizes physical con-

tact; its chemoreceptors give it the ability to taste and smell. There are a few salp cells that are sensitive to light; but there are not a sufficient number of these to enable us to classify the animal as sighted.

That's all the salp has to sense its world. That's about all it needs for the simple life it leads. These animals are usually found in long chains of individuals that move languidly as they pump water through bodies to fulfill their basic needs. Because their requirements are so elementary, whatever elaborate sensory system the salps may once have had has degenerated, until today they have a very rudimentary one.

Salps. Water is pumped in and out of these salps (above), bringing messages from the environment.

A sea cucumber (right) lifts its tentacles perhaps to taste and almost certainly to feed and breathe.

Sea Cucumbers

In and under the skin of the sluggish, slow-moving sea cucumber is a network of nerves, stemming from a nerve ring located near the base of the tentacles. Interspersed with that network are a large number of wartlike projections on the body surface, which house nerves connected with the network. All of these nerve fibers are linked to sensors that receive messages through the water—chemical ones, which the sea cucumber may smell or taste, or tactile ones, which the animal feels. The network in and under the skin is like a mesh of fibers that cover the sea cucumber's body and is most sensitive at the two ends. The sea cucumber's tentacles are also supplied with nerves from the circumoral nerve ring. These apparently are for taste, smell, and controlling tentacle movements. Sea cucumbers also have statocysts which contain inorganic granules that respond to gravity and tell the animal the difference between up and down.

The **green moray eel** is a fairly common resident of the reefs. It hides by day in caves and hunts by night, tracking its prey largely by scent.

The **Nassau grouper** is a fairly sedentary fish. Its olfactory sense is important in providing it with long-range information about its environment.

Detection of Odors

When an odor is released in water, it drifts gently away from its source, scattering like smoke rising from a fire. Then currents pick it up and carry it to more distant areas. Ideally, the currents should be strong enough to carry the odor far from its source but not strong enough to disperse it and make it undetectable to sea life. An animal receiving an odor may react in a variety of ways. It may be attracted to, or driven away from, the source of the odor, depending on the nature of the message it bears. Or it may react by some other nondirective behavior, for example, the settling of larvae.

Man cannot smell anything underwater, but fish can. Why? Our olfactory sensors are located in our noses. When we submerge, air is trapped in our noses; and if water is allowed to penetrate our nostrils, the burning pain due to the difference of salinity between water and human tissues obliterates any sensitivity to odors. Some marine mammals seal off their nostrils with flaps of skin to ensure that no water enters them. In contrast, fish do not breathe through their nostrils, so water may pass through them without affecting their respiratory process. Most fish have one olfactory receptor on each side of the upper portion of the snout. Each receptor is a pit, which is lined with sensitive tissue, folded into a series of ridges and valleys. The folds increase the amount of tissue exposed to the water, without increasing the size of the receptor. The nasal pit is covered by a roof, protecting the delicate tissue inside. One or two holes in the roof admit water.

There are several ways in which water can circulate through the nasal pit. If there are two openings in the roof and the fish is a fairly active swimmer, its normal swimming motions may force the water in one nostril and out the other. The water enters the forward opening, swirls over the sensory tissues, and then exits through the rear opening. In some fish a small ridge at the back side of the front nostril acts as a funnel, directing the flow into the opening. The shortcoming of this type of circulation is that it only works when the fish is swimming or facing a current. A second way in which water is brought into the nasal pit is by a pumping action. The movement of a fish's jaws, forcing water through its gills for breathing, is linked to the nostrils, and the motion draws water into and forces it out of the nasal pit. Fish that are not too active benefit from circulation by pumping, since they can smell even when they are not moving.

Finally, water may be circulated by the cilia, hairlike projections growing inside the nostrils. The beating motion of the cilia is only effective in driving the water through narrow, enclosed spaces. The tubular extension of the forward nostril of eels is ideally suited to this end. Circulation of water by cilia is not fast, but the water in the nasal pit is constantly changing.

The ability of a fish to detect odors is not necessarily related to the size of its olfactory organ. Large organs may improve smelling ability, but in some experiments the fish with the sharpest sense of smell were not those with the largest organs.

Blue angelfish. *The forward and rear openings to the nostrils of the blue angelfish are very close together. The forward one is larger than the rear and allows water to enter the pit easily.*

Electrical Receptors

The ratfish is a member of a group of fishes named chimeras after the fabled monster that had the head of a lion, the body of a goat, and the tail of a serpent.

There are some unusual things about this strange fish that are reminiscent of many different animals. It is related to the primitive sharks and the skates and rays. Like these fish, the ratfish has a skeleton of cartilage. Other fishes have skeletons of bone. It resembles the shark in having a fusiform shape; it is similar to the skates and rays in having a smooth, scaleless skin. But it differs from both of these groups of fish in having a single gill opening on each side. Sharks, skates, and rays have five gill openings on each side (except six- and seven-gill sharks).

Although the sense of electric field percep-

*The **ratfish** (above) lives most of its life in deep water along the coast from southeastern Alaska to southern California. The prominent lateral line branching around its head is a site of ampullae of Lorenzini, which may sense electric fields.*

*The **gray nurse shark** (right) inhabits water around Australia. They are often considered danger-ous. It is thought some sharks use an ability to sense electrical fields for navigation.*

tion has not been studied in ratfish, it would seem likely that these fish have such an ability. On the head of the ratfish are a series of very prominent lateral line extensions which have ampullae of Lorenzini associated with them. In some sharks these ampullae, small jelly-filled sacs, have been found to sense electric fields and possibly serve as a navigational aid.

Sharks sense electrical stimuli like the ratfish—through the numerous pores on their heads that lead to the ampullae of Lorenzini.

Considerable research has been done in the field of electrical reception in fishes with particular emphasis on sharks. It has been found that at least one species of sharks may use this sense as an aid in hunting prey and in approaching their prey to eat them. Whether or not this is true of all the shark species isn't yet known because not all the more than 200 species of sharks have been studied.

Additional research is being done to ascertain whether sharks, skates, rays, and chimeras use their ability to sense minute quantities of electricity in navigation. Many of these fish live in the open ocean (rather than in semienclosed bays and harbors) and some range thousands of miles about the sea. It may be that they have no need to navigate and that it doesn't matter where they are. Yet some studies have showed that a few sharks regularly inhabit a certain area or a particular type of habitat and navigational ability would help them locate the kind of area they prefer.

Abalone

The abalone is a snail with a bowl-shaped shell on its dorsal side and a heavy muscular foot on its ventral side. This muscular foot is in contact with the rocks the abalone lies on and travels on. On the foot and all the other exposed parts of the abalone there are nerve endings, which are extremely sensitive to touch. This is very apparent to the abalone diver. As the diver approaches an abalone, he is preceded by pressure waves set up by his passage through the water. The abalone apparently senses this change of pressure through sensitive nerve endings. To protect itself, the univalved animal clamps itself onto the rock with tremendous suction. An abalone can hold on to the rock beneath it with hundreds of pounds of pressure. It is impossible for a person to remove an abalone that has attached itself this way unless he uses a tool of some sort—divers use abalone irons or a rugged pry bar to collect these animals whose flesh is highly prized.

Nudibranch

Nudibranchs, which are molluscs, are a kind of shell-less snail, often with brilliantly colored cerata on their backs. At the front or anterior end of the nudibranch are one or two pairs of tentacles, sensitive to either touch, smell, or taste. Which of these stimuli the tentacles are sensitive to differs with the species. Some species can be led around by drawing ahead of them a broth made of nudibranch food. This would demonstrate nudibranchs can smell. But in some species the tentacles can only sense the broth if they touch it. This would show this animal cannot smell but can taste with its tentacles. Smell involves the detection of a few molecules free in the medium, whereas taste is limited to the detection of very high concentrations of molecules as on a surface. Consequently touch is often a necessary element for taste. Through the senses of touch and taste, the nudibranch *Tritona* responds to contact with one of its predators.

31

The Touch of Octopods

The octopus has a well-developed nerve ganglion that acts as a brain and a complex nervous system. Since its eye resembles our own, its visual sense is thought to be very good, and its sense of touch appears to be well developed too.

These molluscs are a favorite subject for laboratory experiments. When the food is presented, the animal approaches it directly, traveling in a straight line from its resting place. If a pane of glass is placed between the octopus and the food, the animal doesn't try to get around the obstacle. Instead it stretches an arm around or over the glass in quest of the reward. But if one of its groping arms contacts the food, the octopus immediately slithers around the glass and engulfs the meal. Since the octopus finds its way around the pane of glass only after touching the food, it's apparent that it makes good use of its sense of touch in feeding and in learning about its surroundings.

Curiosity. *Octopods are not social animals, and normally keep their distance from one another. Things are different when the time comes to search for a mate. The male above investigates a female.*

Probing. *With sensitive probing tentacles, the octopus at left climbs over the plexiglass housing of an underwater camera. The suckers on its arms can perceive any irregularities in the surface.*

An acute sense of touch is especially important to the females. They lay their eggs in long strings and attach them to the walls or ceilings of their secluded lairs. The eggs are extremely fragile, and a slight jarring will cause their membranes to burst open. The mother manipulates the strings almost constantly, caressing them with her suckers and tentacles to keep dirt and fungus growths (which would doom the embryos) from accumulating on the eggs. She does this for two or more months (depending on species)

until the eggs hatch. During that time apparently none are damaged, a most remarkable feat considering the amount of time she spends handling them. Were her sense of touch less delicate, this would probably be impossible for her to do.

How do octopods sense by touch? The octopus's eight arms, or tentacles, radiate from its bulbous head and body. Each arm is lined with two rows of cuplike suckers, which are used to grasp objects. The sense of touch is particularly acute in these suckers, and the rims of the cups are especially sensitive. When an octopus encounters an object that arouses its curiosity, it examines it with the suckers, pressing them over its surface. In experiments, blinded octopods, using only their sense of touch, were able to differentiate between objects of various shapes and sizes as well as normally sighted octopods were able to do.

Sensing the Pull of Gravity

Shrimp. Pockets lined with sensory tissue and containing sand enable them to sense the pull of gravity.

Impulses that we are not aware of are continuously being sent to our brains, keeping us informed of our status with respect to the gravitational field of earth. Most animals are sensitive to the pull of gravity. Even those which live in the sea, where the pull of gravity is nearly balanced by the buoyant force of the water, respond to gravity's pull.

In the higher animals the equilibrium organ is in the ear, but some lower animals have special equilibrium organs called statocysts. Impulses are sent to the central nervous system from the statocyst, enabling the animal to tell up from down. An animal may have one, two, or several statocysts. Shrimp and lobsters have two, one at the base of each antennule (the sensors that are adjacent to, and look like a second pair of, antennas).

The statocyst is a chamber filled with fluid and lined with sensitive tissue. Hairs grow from this tissue, projecting into the fluid. Tiny particles, called statoliths, may be calcium carbonate, sand, or pebbles; they rest on the bed of hairs within the statocyst. The statoliths move when the animal moves, since they are constantly being drawn downward by the pull of gravity.

When a shrimp or lobster is standing on the substrate (its normal position), the statoliths press on the bottom of the statocysts. The sensory hairs on which they rest are bent by their weight and trigger nerve impulses to the nerve center. When the animal changes position, the statoliths move, bending other hairs and changing the nerve impulses that inform the animal of gravity.

Shrimps lose the lining of their equilibrium organs, and the statoliths, when they molt. The lining redevelops as the new shell hardens, and the shrimp shovels in grains of sand from the bottom to replace the lost statoliths. This renewal of the statocysts allowed an interesting experiment to be conducted with molting shrimp to show the influence of gravity on their lives. If iron filings are substituted for the sand on the bottom, the shrimp will kick these into its statocysts, and unusual "gravitational" forces can be created by bringing a magnet near the shrimp. The magnet attracts the particles of iron to it in the same way that gravity draws the particles of sand downward. When a magnet is held over the shrimp's head, particles are drawn to the top of the statocyst. The particles disturb the sensory hairs there. The nerve impulses they transmit tell the shrimp it is upside down, so the animal turns over on its back to bring the statoliths back to the bottom of the statocyst. This shows how great an influence gravity has on its life; the shrimp may ignore information provided by its other senses when it rolls over but very often young lobsters remain clinging upside down to the ceilings of their caves.

Awareness of the pull of gravity is more important to the shrimp's relatives, the copepods. These creatures live in the middle depths of the sea, migrating toward the surface in the evening and back to the depths in the morning. For animals with this life-style, the ability to sense the pull of gravity probably helps them travel in the proper direction, which is primarily indicated by light. The light is of such low intensity, however, additional sensitivity may be necessary.

*This **Pacific lobster** emerging from a hole in the rocks changes its statolith each time it molts.*

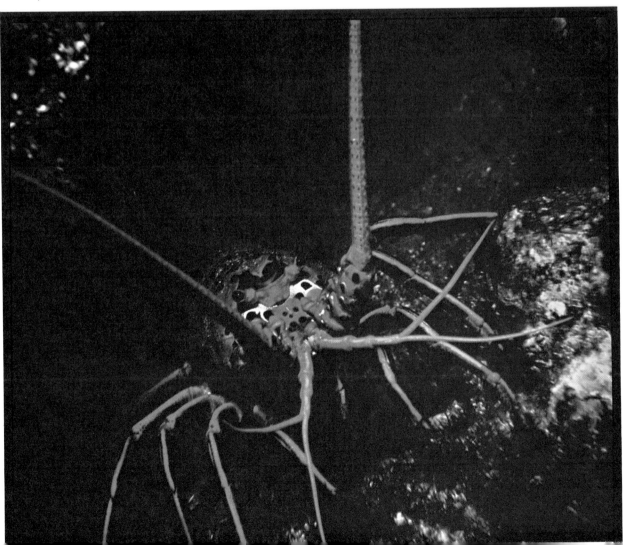

How the Ocean's Temperature Affects Fish

Fish live nearly everywhere on earth, wherever there is water, from warm tropical seas to frigid polar oceans. In the Persian Gulf water temperature readings of 96° F. have been taken, while the water in polar regions can reach a below-freezing temperature of 28° F. These differences are of great significance to animals living in the sea. Fish are believed to sense water temperature with free nerve endings in their skin. These sen-

sors are so refined that they can detect variations in temperature less than 0.1° F.

Some fish (stenotherms) are more critically dependent upon a given temperature than others (eurytherms) which can adapt to sizable water temperature variations. In any case, no fish can maintain a constant body temperature as warm-blooded mammals or birds can, and their life processes accelerate in warm temperatures and slow down in water that is cold.

Those marine creatures that lead a sedentary life in one given area have to be able to cope with the seasonal temperature changes, and are eurytherm. Stenothermal animals, on the contrary, have no other choice than to migrate, vertically or horizontally, daily or seasonally, or occasionally to follow the masses of warm or cold water that best suit them. There are striking exceptions to these rules, like tuna migrating for other reasons, or the polar creatures living actively at the very frontier of life, in areas where their body fluids come very close to freezing.

Water temperature has other influences on the lives of marine animals. It regulates spawning habits and is a controlling factor in the hatching of eggs and growth rates of animals. Knowledge of the tolerance levels and preferences of marine animals will be helpful to man in two ways. First, we will be able to predict where fish will congregate at given times of the year. We will be able to locate schools of migrating fish by using thermometers. Precision temperature readings will make our fishing industry far more efficient than it is at the present time. The second, and more important way knowledge of temperature preferences of fish will aid man is in aquaculture. Aquaculturists will be able to speed up both reproductive and growth rates by maintaining optimum temperatures for fish in fish farms. Heated water from atomic power plants has been used in some cases.

Jacks (opposite page), like all fish, are sensitive to changes in water temperature. Some fish of temperate waters migrate north or south with the seasonal changes of water temperature.

Measuring temperature profiles. At the right, a torpedolike bathythermograph (BT) is lowered over the side. The BT makes a continuous recording of temperature changes relative to depth.

Lateral Line–The Sixth Sense

Garden eels live in tunnels in the sand. When they reach out, their lateral line organs, appearing as evenly spaced dots behind their heads, are exposed to a barrage of messages carried by pressure waves.

Garden eels probably detect approaching danger with their eyes and their lateral line system, a sixth sense peculiar to fish and to amphibians in their aquatic stage. A swimming fish produces low-frequency vibrations (referred to here as pressure waves) which can be received by itself and others in a school at very great distances and which echo back from objects, helping the fish to orient itself. The lateral line, which appears as a line or series of dots running down the sides of the fish, picks up such signals or echoes. It is probably one of the most important senses fish have. Some predators use it to detect prey far beyond sight range. Since it receives pressure waves, this sense may in some ways be considered both distant touch and hearing.

How the Lateral Line Works

The lateral line sense organs of fish are similar in design to those of its inner ear. They are contained in canals, which are stretched out along each side of the fish's body, and branches of which encircle the fish's head. The canals are located just beneath the skin, and pores through the skin or scales open them to the water. In some fish these pores are quite large and visible, like those of the garden eels on the facing page. But in others they may be very small and difficult to see.

The basic components of the lateral line system are the neuromasts that function in the same way as the cristae of the inner ear. These are collections of sensory cells, each with tiny hairs projecting from it. When a disturbance near the fish sets up a pressure wave, the moving water strikes the fish and disturbs the mucus in the pores and canal. The mucus, in turn, jiggles the hairs of the neuromasts. The hairs stimulate nerves, triggering the discharge of nervous impulses to the brain and conveying low-frequency messages that have traveled great distances.

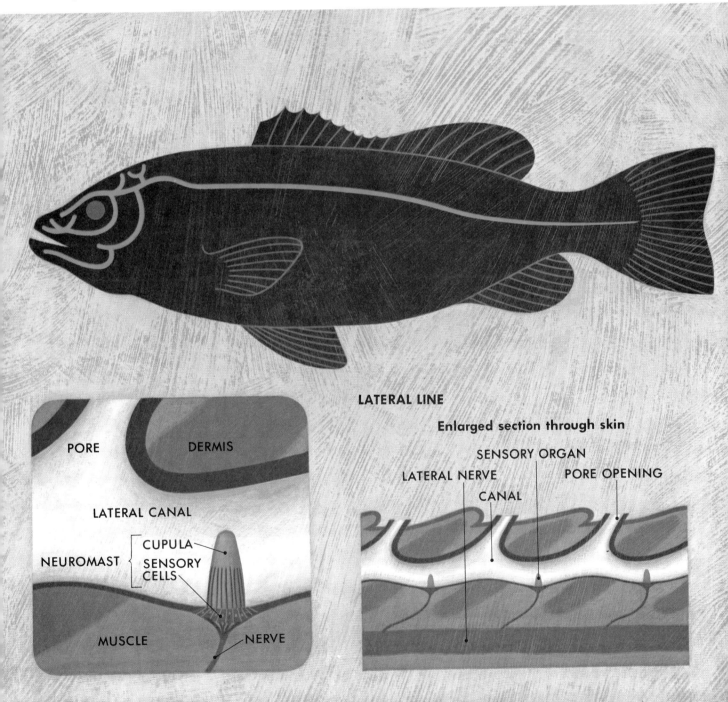

PORE   DERMIS

LATERAL CANAL

NEUROMAST   { CUPULA   SENSORY CELLS

MUSCLE   NERVE

LATERAL LINE

Enlarged section through skin

SENSORY ORGAN
LATERAL NERVE   PORE OPENING
CANAL

Triggerfish responds to signal. This triggerfish regularly responded to a tapping at the porthole by oceanauts inside Conshelf II.

Responsive Triggerfish

Less than a year after two divers spent a week in Conshelf I, the first manned undersea station, a team of five divers descended to 35 feet to live on the sea floor in Conshelf II. Their stay lasted a month, and during that time they had an opportunity to observe the reef life around them as no one

else ever had. The oceanauts lived in the environment, instead of just visiting it for a few hours at a time.

The men could work outside the habitat, called Starfish House, for six hours a day. During their stay the cook, Pierre Guibert, tamed a triggerfish. Sometimes during his swims along the reef Pierre opened the shells of tridacna clams and then fed the clams to

*This **triggerfish** became especially fond of eating tridacna clams "on the half shell" during the Conshelf II experiment in the Red Sea.*

the triggerfish. After a while it seemed as if the triggerfish recognized the cook, since it constantly went to him, even though all the oceanauts were dressed alike. Since the men were covered from head to toe by their wetsuits and masks, how did the fish distinguish among them? How did it sense which of the divers was Pierre? Perhaps the rhythm of Pierre's breathing set him apart, or perhaps the bubbles popped differently from those of the other divers. After meals Pierre tapped his ring against the inside of one of the plastic portholes. The fish would glide by the window and look in to see who was making the sound. When the fish was sure that Pierre was the source of the signal, it swam right to the open entry hatch to get food scraps.

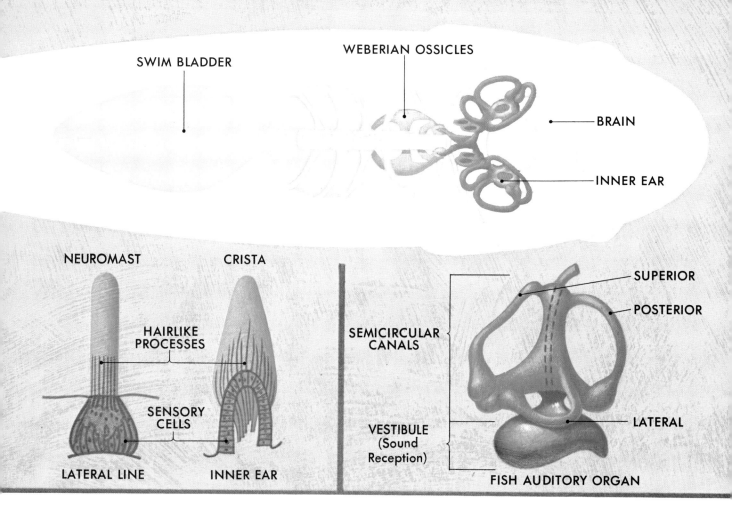

SWIM BLADDER

WEBERIAN OSSICLES

BRAIN

INNER EAR

NEUROMAST

CRISTA

HAIRLIKE
PROCESSES

SENSORY
CELLS

LATERAL LINE

INNER EAR

SEMICIRCULAR
CANALS

SUPERIOR

POSTERIOR

LATERAL

VESTIBULE
(Sound
Reception)

FISH AUDITORY ORGAN

The Ears of Fish and Man

The ear of both fish and man serves two functions, to maintain equilibrium and to perceive sounds. A human ear is a three-part organ, consisting of an outer, middle, and inner ear. Fish ears are more simple and are made up of only an inner ear. The inner ear in both man and fish is where sounds are actually sensed. It's made up of a fluid-filled chamber designed to perceive sound vibrations, gravity, and motion.

When sound reaches man's pinna, the fleshy, cone-shaped part of the outer ear, it is channeled into the auditory canal, the tube leading to the eardrum. When the sound strikes the eardrum (tympanic membrane), vibrations result. A series of three bones (auditory ossicles) relays the vibrations of the tympanic membrane across the middle ear to the

inner ear. These bones act as levers; their action reduces the amplitude of the vibrations, while increasing their force by a factor of 22. Thus the force applied to the inner ear is 22 times greater than the force causing the tympanic membrane to vibrate. This is very important, since without amplification the vibrations of faint sounds would be too weak for us to hear. The movement of the auditory ossicles disturbs the fluid in the canals of the spiral-shaped cochlea of the inner ear, bending its tiny sensory hairs and triggering nerve impulses to the brain.

Fish, like man, have two ears, called labyrinths. We cannot see ears of fish, since they are enclosed in the fish's head and have no external openings. (Sharks and rays are an exception; a small tube leads from their ears to the water.) The ears are found on either side of the fish's head, just behind its eyes. They have no need of a pinna like man's, be-

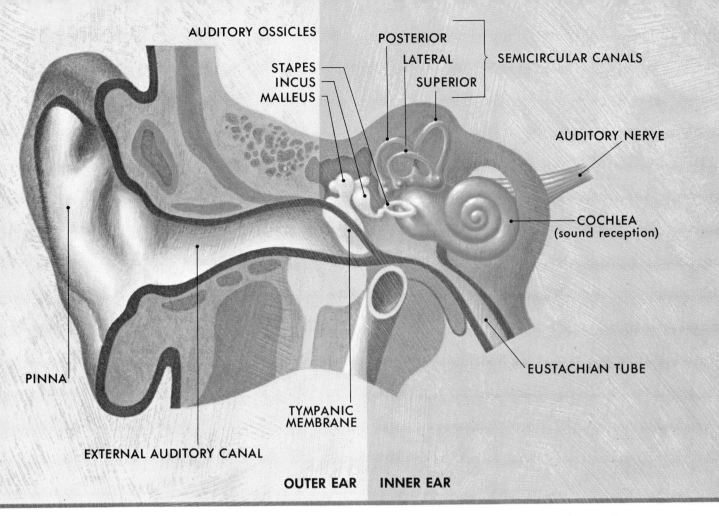

AUDITORY OSSICLES

STAPES
INCUS
MALLEUS

POSTERIOR
LATERAL
SUPERIOR

SEMICIRCULAR CANALS

AUDITORY NERVE

COCHLEA
(sound reception)

PINNA

EUSTACHIAN TUBE

TYMPANIC
MEMBRANE

EXTERNAL AUDITORY CANAL

OUTER EAR   **INNER EAR**

cause fish are "acoustically transparent"—sound vibrations pass through them as they do through water, and strike the ears along the way.

The organs of hearing and balance are basically similar, being fluid-filled chambers in which sensitive hairlike cells respond to movement of the fluid within the chamber. Because of this relationship and evidence from primitive fish, it is thought that the sense of hearing developed from organs of balance. In both man and fish the three semicircular canals detect any turning motion through a movement of fluid within the canals, each of which corresponds to one of the three dimensions. The fluid motion in turn stimulates sensory hair cells which initiate nervous impulses to the brain. In fish these hairs are imbedded within a matrix and together constitute a crista. Similarly the organs of vibration reception (neuromasts) in the lateral line of fish are made up of sensory hairs imbedded in a special substance which is affected by fluid motion.

The organs for hearing in fish and man differ somewhat but maintain the same basic elements as seen in the semicircular canals. In fish two fluid-filled vesicles each possess a calcareous stone, called an otolith, which rests on sensory hairs. The motion that sound imparts to the fluid moves the hairs relative to the otolith, thereby stimulating them. Instead of the calcareous earstone seen in fish, man possesses a tectorial membrane with which sensory hair cells are in contact. Sound creates a motion in the fluid which moves the tectorial membrane and in turn stimulates the hair cells. In some fish hearing is enhanced by the use of the swim bladder in a manner analogous to the ear drum and by modified vertebrae (Weberian ossicles) which act like the auditory ossicles in man.

Chapter III. Man's Mechanical Sensors

Although noises do exist underwater and travel clear, fast, and far, to the naked ear of a diver they sound extremely muffled, like those of a city blanketed by a heavy snowfall. To hear underwater sounds, a diver must hold his breath, or the sound of his exhaust bubbles will overpower them. Under the best conditions divers can perceive the sound of a bell several miles away or a small explosion 100 miles away. Even so, air bubbles in the canal of the outer ear and the air space of the middle ear block out many sounds, and marine animals are much better equipped for hearing underwater.

Because we are strictly terrestrial creatures, we are not particularly sensitive to some of the waves of information coursing through the water. We can tell if water is cold or hot or if it is salty or fresh, and we can hear sharp underwater noises. To overcome our insensitivity we have devised mechanical sensors to do the job for us.

Medium- and low-frequency sounds travel well through the ocean and, depending on the wavelength of the sound and the size of the object, are reflected by any object conducting sound at a different velocity than water. Consequently echo sounding has become a primary tool for the exploration of the oceans. Sound-producing transducers transmit a pulsed signal, and hydrophones receive the returned signal. The information provided by these echoes has given us some insight into the vertical migrations and the schooling habits of some marine creatures, and has revealed a jagged submarine landscape practically free of erosion, but partially buried under miles of sediment.

Magnetometers measure the earth's magnetic field; gravity meters measure its gravitational attraction. They are useful tools in the quest to learn more about the structure

and the history of our planet. Such information tells of changes that have occurred on earth since the ocean basins and continents were formed and of changes that are occurring now. Soon such data will enable us to predict eruptions, earthquakes, landslides, and tsunamis. Our senses cannot precisely assess the temperature of seawater or its chemical makeup. These characteristics affect the density of the water and consequently the way sound travels through it, as well as the dynamics of the water masses.

Chemical analysis of seawater reveals the relative concentrations of the many elements dissolved in the oceans. Chemicals in suspension in our atmosphere directly influence the life of air-breathing plants and animals. In a similar way, salts and minerals

> "Man is still forced to rely on a variety of instruments to supplement his natural sensors."

dissolved in water are directly in communication with fish blood through the thin membranes of gills. Sea mammals, on the contrary, have only an indirect connection to the sea's chemistry through the digestive system. And man, even more isolated, has to rely on instruments to supplement natural sensors. The information gained from a better chemical monitoring of the sea may help us to learn more about our planet, as well as to exploit the oceans by ocean farming, for example, without destroying them or their populations.

Hydrophones pick up underwater sounds, transmit them to an amplifier and then to man's ears. Here, in the Florida channels, Captain Cousteau observes baby alligators alternately diving and surfacing to crawl on the water hyacinths.

Reading the Ocean Floor

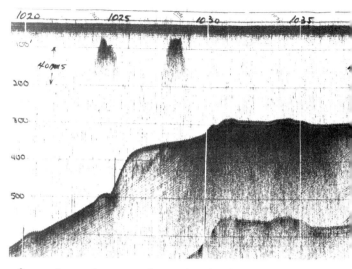

Invisible messages can teach us about the ocean and the topography of its floor. Some of these messages are the echoes of sound beamed at the sea floor. Others are hidden in visible light.

Echo sounders and sonars use sound waves to study the ocean. These instruments have three basic parts—a transducer, aimed vertically in echo sounders or obliquely in sonars; hydrophones; and a timer. Most of the time the transducer can be used as the hydrophone, which simplifies the instrument. The transducer vibrates when it receives an electrical impulse. These vibrations generate sound waves in water, which travel in the main direction the transducer is pointed. When the sound strikes a target, it is echoed back to the hydrophones, also mounted below the ship's waterline. The hydrophones sense the echoes and translate them into electric impulses connected to the stylus of a paper recorder. Sometimes the information is registered on a cathode tube like a television screen in place of the paper. The timing device indicates how long the sound took to go from the transducer to the reflecting object and back to the hydrophones. This length of time tells us the water depth in the case of an echo sounder or the location

of a submarine or of a school of fish when sonar is being used.

The messages hidden in visible light are uncovered by a technique called multispectral imagery. In this technique, a color photograph is taken of a section of the sea from aircraft or satellites. The picture can tell us about the topography of the land and the shallow sea floor. It may tell us about temperature gradients. It may reveal varying densities of the seawater. The differing populations of phytoplankton in the sea may be shown. The spread of thermal pollution may be demonstrated. Soon multispectral imagery from manned orbital laboratories will monitor the oceans and give us information about its productivity or its pollution.

Seismic profile above, which covers 1.4 nautical miles of the shallow Mediterranean continental shelf off Monaco, shows different layers of sediment. The dropoff, or "knee," at about 360 feet may be an ancient shoreline.

*A **sonar crewman**, left, monitors the echo-sounder tracings as impulses are recorded. Such tracings give a profile of the ocean floor over which the vessel carrying the sounder passes.*

Multispectral photography. *Variations in water depth, temperature, plant life, topography, and water turbidity are some of the messages revealed in the multispectral photograph at right.*

47

Towing device houses submersible sonar, which performs better than a surface-stationed sonar.

Magnetometer is ready to be lowered over side of a research vessel to measure earth's magnetic field.

Oceanographic Instruments

Man's senses are too inadequate to perceive, much less measure, all the invisible messages available in the sea. To study such a wide array of phenomena, man has devised a multitude of instruments to help in observations, measurements, and recordings. He may do something as simple as pouring a dye into the sea to observe and analyze a current. Or he may use the most elaborate sensor devices to send and receive electronic impulses for fine measurements of many kinds. Each discipline within the realm of oceanography has its own specialized instruments. But because the various branches of oceanography overlap each other to a considerable degree, some instruments are used by most oceanographic scientists. But each researcher may not need the same accuracy for the same information. Thus a biological oceanographer uses a salinometer to determine at what salinity certain animals are living. The chemical oceanographer wants to know the salinity to control the chemical makeup of the water. The geologist is interested in it to understand the effect of the substrate on the

water. And the physicist wants to compute the general displacement of water masses or the velocity of sound.

Some instruments, basically similar to those used on land, need various adaptations for use in the sea. In measuring the temperature of ocean depths, for example, a thermometer cannot simply be lowered on a line to a predetermined depth. When it is retrieved, it must pass through warmer waters near the surface, and the reading on the thermometer will change. To overcome this problem, the thermometer's mercury column must be fixed in position before retrieval. So the instrument is attached to a revolving mount, generally including a water sampling bottle, and lowered. When it is ready to be brought back to the surface, a heavy metal plug called a messenger is slid down the line. It strikes the top of the thermometer housing and releases a catch; the thermometer is freed from the line at that end, while its other end remains attached to the line. The upper end, now unattached, sinks below the bottom end, and the thermometer is turned upside down. The mercury column is broken and isolated and the thermometer can be

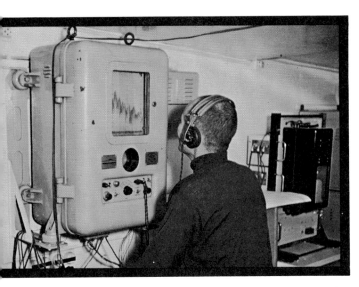

Echo sounder. *Specialist monitors echo sounder, watching as a profile of the bottom is recorded.*

retrieved without changing the reading for the predetermined depth. Variations and improvements on the thermometer have resulted in many refinements and in another instrument called the bathythermograph, or BT. The BT directly traces, on a small piece of smoked glass, while it is lowered, the curve of temperature versus depth.

Electronic instruments like thermistors have been developed to replace the traditional mercury thermometers or to determine the chemical makeup of water samples. By dipping electrodes into a water sample, we can now determine the amount of oxygen dissolved in the sample. Complete chemical analysis of water samples are now routine procedures, completely automated, in modern laboratories. Another device senses the infrared radiations from the surface of the sea, which are directly related to the temperature of the thin molecular layer at the surface, where at the same time take place absorption from the sun's energy and evaporation of water into the atmosphere. Thanks to remote sensors, the exact temperature of this critical zone can be monitored quickly by aircraft or even by satellites. It allows accurate mapping of the ever-changing frontiers of the ocean currents and it helps in meteorological forecasts. Many chemical tests can be performed quickly and accurately from samples with the help of an instrument called a spectrophotometer. This very short roundup is extremely incomplete. New devices that can gather more data are constantly being developed; they will assist all types of oceanographers, whether they work from vessels or aircraft or through satellites. All these figures are poured into huge computers in the hope that a global understanding will ultimately emerge from this jungle of information.

In the 100 years since the expeditions of H.M.S. *Challenger* gave oceanography recognition as a science, many new instruments have been invented. Many of the early instruments were simple, basic tools. And many of today's instruments like water and bottom samplers, current meters, and corers are nothing more than elaborations on those early devices. However complex, however ingenious, most of these exploration tools fail to give us a general knowledge of the oceans; we are studying a giant with Lilliputian goggles. A complete revolution in our approach to studying the sea is shaping up: large-scale rotating models of the oceans allow us to reproduce and study the mechanisms that produce currents; networks of anchored buoys, fully instrumented, transmit data continuously, and more and more aircraft and satellites will enable us to reach the ultimate scale of investigation that the oceans deserve: the global scale.

But whether the new tools of the oceanographers are complex or simple, whether their field is local or global, they all serve the same end—to translate or use invisible messages of the sea and increase man's knowledge of the world we live in.

Satellites Study the Oceans

For thousands of years man had to be content with studying only small areas of the oceans from the shore or from the deck of a ship. But with the coming of airplanes, observations and photographs of larger areas of the ocean's surface became possible. However, scientists still wanted, and needed, an even more encompassing look at the earth. A breakthrough came in 1957 when Russian scientists launched the first man-made satellite. This began a new era of ocean (and terrestrial) exploration. Orbiting high above the earth's surface, satellite cameras photograph thousands of square miles in a single exposure. Sequential photographs taken only hours apart are pieced together to show huge areas of the earth's surface. This enabled scientists to see relationships and interactions between land, weather, and water masses—relationships that might never have become evident from surface observations alone. Devices other than cameras are carried aboard some satellites to measure physical characteristics of the sea more easily and accurately than we can from the surface. These characteristics include surface conditions over large areas, temperature, and spread of pollutants.

Pictures of the ocean's surface give an interesting and informative look at shallow submarine terrain. Subtle contours, textures, and color variations show up with striking clarity. Changes in bottom topography, brought about by currents or rivers spilling into the oceans, and shallow bays can be seen. By comparing photos taken over a period of time, changes that have occurred in the shape of land masses by erosion or deposition of sediments are seen, and specialists may be able to predict changes that are likely to occur in the future.

Satellite photographs and measurements

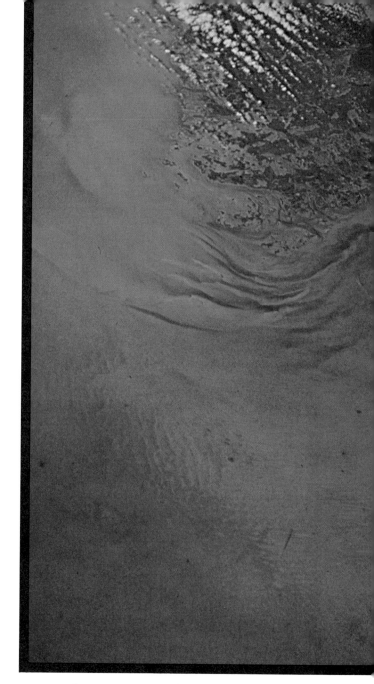

are used to support the theory of continental drift by revealing startling similarities in the geological formations and shoreline contours where the land masses were thought to have been connected millions of years ago.

Sailors, fishermen, and others living or working upon the seas receive great benefit from satellite observations. Satellites measuring roughness of the surface waters, monitoring weather systems, detecting changes in polar ice packs, and tracking the movement of ice-

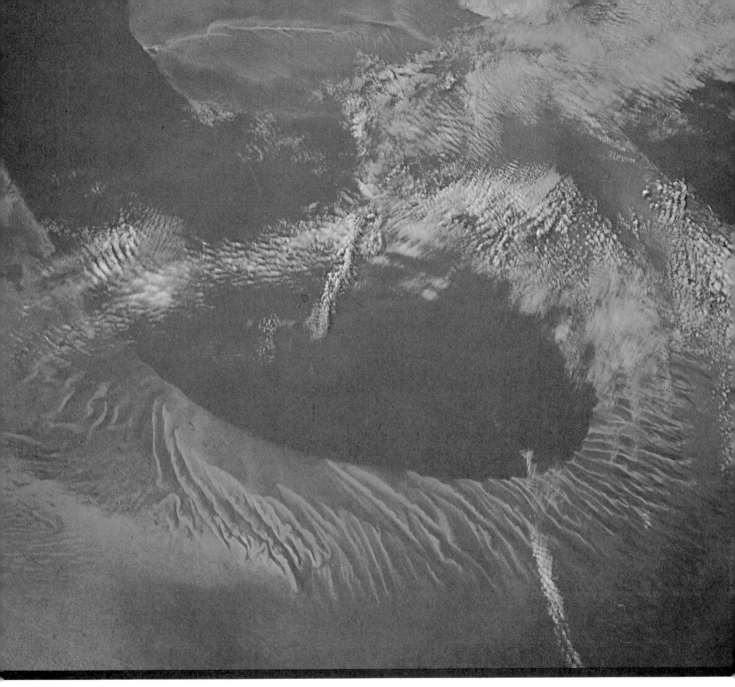

*In this **satellite photograph** of the Bahama Islands, the deep-blue Tongue of the Ocean is like a gouge in the coral sand of the ocean bottom. The shoal waters are only 35 to 40 feet deep. Farther out, the bottom drops to a depth of 4400 feet.*

bergs, can warn in advance of hurricanes, storms and other dangerous situations.

By using infrared photography, satellite photographs can be taken by night as well as by day and show the temperature of surface waters. The limits of such currents as the Gulf Stream or the Humboldt can be clearly read on these pictures, as well as areas of upwelling of cold bottom waters. Unfortunately infrared pictures do not penetrate cloud formations and only one-third of the ocean's surface can be photographed be-

tween clouds. New microwave techniques will soon be able to give us similar information through the clouds.

Oceanography by satellite opens many doors for scientists. Its prospects are limited only by our imagination and technology.

51

Instrument Packs

How can we learn the secrets of the dolphin and whale? Going with them would probably be the best way. But because we can't do that, the next best approach is to send along a mechanical emissary.

We'd like to know more about the everyday life of these animals. We would like to know more about how deep they dive, how long they can hold their breaths, how they find food, how they communicate with each other, how they navigate from one part of the world ocean to another, how they behave and how their behavior can be modified, how the various organs of their bodies work. In fact, we'd like to know all there is to know about these intelligent sea creatures.

In the United States scientists have been studying cetaceans for years seeking answers to some of these questions. One method has been to attach instruments to the animals and turn them loose in the area. The instruments, equipped with tiny radio transmitters, send information back each time the animals surface for air. They tell of water depth and properties as well as the animal's location. The instruments are attached with a material that will corrode when the battery runs down. So when the power supply is used up, the instrument pack falls off, leaving the animal unfettered. Studies showed that the device increased the drag by only 2 percent as compared with unhindered swimming dolphins, and showed no increase in respiratory rate. Unfortunately the manipulated dolphins are psychologically disturbed, and they rarely rejoin their pack.

*A **common dolphin** swims at the surface with its newly attached radio transmitter mounted on its back around its dorsal fin. The white rod is an antenna, which enhances radio transmissions.*

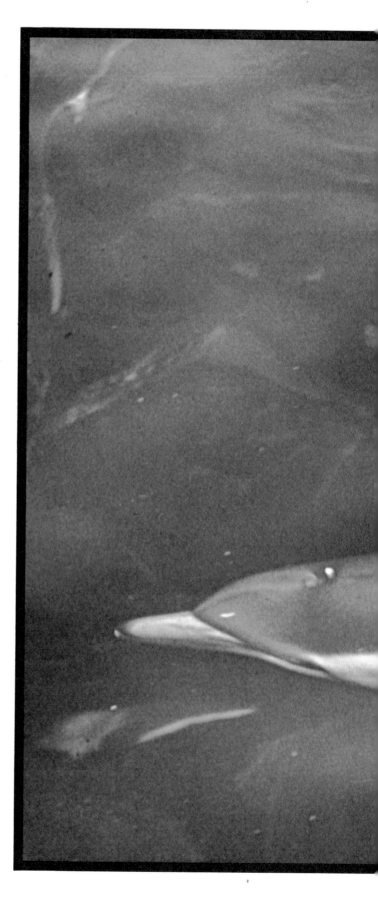

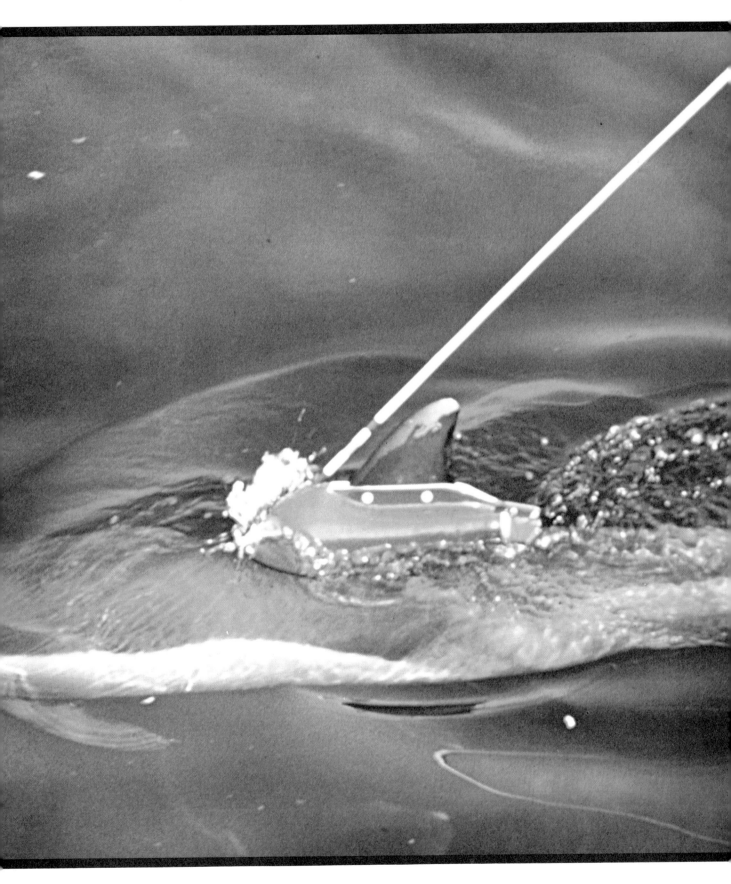

Chapter IV. Voices and Drums

In the sea a variety of sounds are broadcast by a multitude of animals. The most frequent and prolific sound producers come from three groups of animals—fish, crustaceans, and mammals. Just how, when, and why they produce various and wide-ranging sounds is being studied by many scientists. Thus far, these studies have told us only a small part of the whole story of sound production in the sea. Scientists have found that no fish, crustacean, or other invertebrate animal has vocal cords to help it produce sounds in the same manner we do. But they use a number of other methods to do so.

Some of the sounds in the sea are produced by rubbing or grinding together hard parts of the animal's body. This is known as stridu-

> "Studies have told us only a small part of the whole story of sound production in the sea."

lation. Stridulation may be accomplished by grinding together the teeth in the jawbones or the pharyngeal teeth that some fish have deep in their throats. Or it can be accomplished by rubbing chitinous claws or other hard body parts together. Or something akin to cracking the knuckles or snapping the fingers may make a sound.

Some fish produce sound with their swim bladder, the gas-filled, balloonlike organ, which also serves as a hydrostatic device to regulate buoyancy. These fish make noise by rubbing or thumping muscles or bones against the bladder. The bladder can also be used to amplify a stridulatory process. It is thought that some sea mammals constrict their larynx or other air passage and force air through it to produce whistling sounds. Parrotfish, some surgeonfish, and some trigger-

fish produce a rasping sound as their teeth scrape algae off coral reefs.

The cetaceans produce several types of sounds. They may click, grunt, whistle, tick, sigh, wheeze, cluck, or rumble. Dolphins and humpback whales may emit sounds and repeat them a varying number of times. Or they may produce a series of phrases in different sequences. These various sounds are referred to as phonation, and the meaning of some of them is clear. But some, like the sounds produced for use in echo location, are still being studied.

It is thought that dolphins produce their sounds by forcing air through constricted passageways in their nasal and laryngeal regions. Air escaping out of the animal's blowhole may be at the origin of whistles and squeaks. They also produce sounds by lobtailing and by clacking their jaws together.

Sea creatures often make sounds in response to stimuli such as hunger, fear, light, anger, or aggression. Some stimuli, like hunger or sunlight, appear regularly, and therefore the sounds related to them have a certain predictability. Some studies have revealed a distinct daily cycle of the sounds caused by different organisms. Other cycles are annual and associated with mating season.

Sea creatures may have many reasons for making sound. Noises may be used for species or sex recognition, for navigation, for schooling, as a threat, or to express alarm.

Sounds from below. When parrotfish bite off and grind chunks of coral rock and other stony materials, the action of their teeth produces loud, harsh sounds. Here a rare African parrotfish bites into calcareous algae atop a rock.

Grunts, Croaks, and Bumps

The Mediterranean croaker above swims past in its usual shallow-water habitat.

French grunts at right school on a coral reef during the day, disperse at night to feed.

Fish of the grunt family get their name from the gruntlike sounds they emit when grinding their upper and lower pharyngeal teeth, far down in their throats. Their swim bladders act as sounding boxes, resonating and amplifying these sounds. But the sounds are not easily heard without hydrophones for underwater listening. The grunting noise, however, can be heard distinctly when the fish are caught and removed from the water. Apparently the sound is a distress signal in such a situation. But since they produce the grunts in their natural habitat, in and around coral reefs, the sound may have different meanings at other times.

Croakers are even noisier than the grunts. Attached to the sides of their elaborate swim bladders are several strong muscles, which the croaker can vibrate much the way a guitar string vibrates when plucked. And in much the same way as the sounding box of the guitar amplifies the sound of a plucked string, so does the swim bladder amplify the noise of the vibrating muscle. The sounds these fish produce in this way are those of croaking or bumping.

The croakers of some species might be particularly noisy early in the evening and reach crescendo by midnight, then taper off. Croakers were found to produce more sound as spawning season approached. The noise level decreased after spawning took place. Croakers seem to have a noise schedule.

Spotted drum. This is a fish that makes sounds by vibrating strong muscles in its swim bladder.

Drums

Drums are the best known of the sound-producing fish. These fish live along warm sandy shores and at times produce nocturnal concerts. Like the croakers mentioned on page 56, almost all of the members of the drum family have swim bladders, which serve as resonating chambers for the sounds produced by the vibrations of strong muscles attached to the bladder walls. These muscles vibrate about 24 times per second. Drums can produce sounds at will, giving off noises of varying pitch—from deep, drum-like thumps to higher-pitched sounds. While the exact reason for them is unknown, sounds do increase during mating season. Then drums seem to be singing in chorus.

Toadfish. *This very vocal toadfish is a member of a family with a wide repertory of sounds.*

Toadfish

The toadfish is a very vocal fish capable of producing two types of sounds—grunts and boat whistles. Grunts are made mostly by males and are most frequently used when two aggressive males are within viewing distance of one another. This sound is modulated by the toadfish depending on how antagonistic it feels and can be intensified into something close to a growl. The boat whistle (also called the foghorn) sound is a low-frequency burst of tone which lasts about one-half second. These sounds are emitted more frequently during the reproductive season. Highly competitive males are more vocal when a female toadfish passes by the male's nest.

Scrawled filefish. This animal's rough skin, once used as sandpaper, gave it its name.

Filefish

Tube-mouthed filefish inhabit tropical waters throughout the world. They are related to the triggerfish and are equally prodigious noisemakers. The sounds they produce are variously described as "the tearing of canvas," grinding, grating, grunting, rasping, and scratchy clicks.

Filefish have two sets of teeth. One set, which grows from the jaws, are incisors that protrude from the tiny mouth of the fish. Another set is deep in the animal's throat; they are pharyngeal teeth used for grinding and crushing. A filefish grinding its incisors against each other produces a metallic rasping. The pharyngeal teeth working to crush shells make grinding and grating sounds.

Porcupinefish. *An inflated porcupinefish drifts helplessly, its body spines erect and fins flailing.*

Porcupinefish

Puffers, porcupinefish, and burrfish produce scraping and thumping sounds, principally when they are inflated. When they are deflated, most of this group of fishes can't articulate the sound-producing parts. So they remain silent. But when they have been threatened and have inflated themselves, the plate in the lower part of their beaklike mouths is thrust forward and can reach the upper plate. When these plates are ground against each other, stridulatory sounds result. The sounds don't come through as loudly as those of other fish that grind their teeth and jaws together. The reason is that these fish have small swim bladders that cannot act as a sounding box.

Mussels and Clams

Mussels and a few species of clams secrete byssal threads to anchor themselves to rocks and to each other. To move, they must break their strong mooring lines. The snapping of byssal threads of the blue mussel, common in the waters of New England on the east coast of the United States, recently has been recorded. The resulting sound— usually between 1000 and 4000 hertz per second (a unit of frequency equal to a complete movement per second)—is a crackling noise. It is the primary background crackling sound produced outside tropical and near-tropical temperate waters.

Mussels are bivalves, related to clams, scallops, and oysters. The tough byssal threads enable them to withstand the force of breaking waves, and in some cases this allows them

Mussels. Barnacle-encrusted mussels (above) lie on the sea floor in the tidal zone, attached to the bottom and to each other by byssal threads.

Clams. Some clams can produce sounds by clapping their valves together (right). Others produce crackling sounds by stretching their byssus to the breaking point.

to live out of the reach of predatory starfish that cannot cling to such tortured surfaces. The byssus starts out as a viscous material secreted from between the clam's or mussel's valves. The secretion runs down a groove in the animal's foot to the cup-shaped terminal end of it which is planted on a rock or some other object. Exposed to seawater, the viscous protein sets to a tough, flexible consistency. The mussel moves its foot and secretes another byssus and another and another until it is firmly anchored by many such threads to its new locale.

A pair of pistol shrimp on a red sponge watch for prey they will "shoot" with their outsized claws.

The Poppin' Pistol

Pistol shrimp have been called the gunmen of the reefs and tide pools. The pistollike sound of these one- to two-inch-long shrimp is made by clapping two parts of the large claw together. The small part of the claw is held perpendicular to the large part of it. When the small part, like a thumb on a mit-ten, is closed onto the large part, or palm of the mitten, it snaps over a ridge. This produces the sharp report that can stun a small animal nearby by the concussion it generates. Pistol shrimp use their "guns" to defend the burrows they live in and to stun their prey. If they lose the large claw, the small one grows up to replace it while the missing one is regenerated.

A tropical spiny lobster warily watches for danger, ready to retreat into a crevice in the reef.

Vibrating Antennas

Divers who have hand caught spiny lobsters in the tropic waters of the Caribbean have felt the vibrations produced by their stridulatory or noise-making apparatus. These noise-making devices are located at the base of each of the lobster's two large antennas. The antennas themselves pick up sounds as they vibrate through the water. They are covered with tiny sharp spines and also serve as defensive weapons. When the diver first grabs the lobster, the animal's stridulatory mechanism starts vibrating and can be felt even through the diver's gloves. The diver may also be able to hear the sound, which is audible to most aquatic animals. The sound may be meant to frighten predators.

Stridulatory Sounds

Most damselfish are small and inhabit tropical coral reef areas. But one member of the family, known as the garibaldi or ocean goldfish, grows larger and lives in the temperate waters of the southern California coast. It is noted for its vigorous territoriality; it is also a prolific sound-producer. The thumps and clicks and scrapes that emanate from the garibaldi are stridulatory sounds, made when the fish grinds its pharyngeal teeth and resonates the resulting sound off the swim bladder. The sounds are made especially when the garibaldi is defending its territory, when it is competing for food, and when it is engaged in courtship and sexual activities. The sounds they produce and possibly their bright coloration are often successful territorial defenses. The brilliant coloration may serve as a warning to intruders in the kelp beds where they live.

In most fish stridulatory sounds result from the grinding of pharyngeal teeth and resonance off swim bladders. But seahorses (and their near relatives, the pipefish) produce stridulatory sounds by a different means. The clicks they make are produced when the seahorse pulls its head up sharply and moves its tiny mouth and jaws. The sound results from rubbing the edge of a skull plate against another bone, which articulates with the jawbone. Like other fish, seahorses and pipefish use their swim bladder to amplify sound. Their swim bladder is a large organ taking up a fair amount of space in the body cavity and running the full length of the body. The sounds are especially loud when the seahorse is feeding and during courtship and mating. It may respond with this sort of sound-producing movement when it takes up a new location in the eelgrass beds it lives in. The sounds are thought to aid the seahorse in orienting itself.

Besides the variations in the types of sounds fishes produce—the scrapes, bumps, rumbles, clicks, and staccatos—the sounds themselves may be variable in length and may be given at different intervals. All these variables give the fish a sizable number of messages it could send, if, in fact, that is what the sounds are produced for. Evidence does indicate that fish use specific sounds for specific purposes—to signal aggressive intent, or in courtship, for instance. Their messages must, however, be of only a very general nature and cannot be considered a language in our sense of the term.

*A slender **seahorse** (right) twines its prehensile tail around a blade of eelgrass. The sounds these little fish produce when they are moved to a new location are thought to help them orient themselves.*

Garibaldi. *This colorful large damselfish (left) lives in kelp forests off the coast of California. It is highly territorial and very vocal when trespassers come within its invisible property lines.*

*A **squirrelfish** displays its spiny dorsal fin while it produces grating sounds to ward off invaders.*

Strumming of Muscles

The wide assortment of sounds squirrelfish make have been described as knocks, staccatos, grunts, and growls. Most of these sounds are made by strumming a large muscle against the gas-filled swim bladder. In squirrelfish the swim bladder gives the fish a relatively large sounding board to resonate the strumming muscles. In addition, the swim bladder is supported by the first three pairs of ribs. These may contribute to sound production by being moved in and out. Squirrelfish also grind their pharyngeal teeth to produce sounds, which are amplified by the nearby swim bladder.

Snappers have the ability to produce sounds with the swim bladder and associated muscles.

Multitude of Sounds

The thin-walled swim bladder of the snapper is enclosed within a tough membrane. To produce sounds, most of the members of this large family of fish vibrate muscle fibers against the strengthened swim bladder by rapid muscular expansions and contractions.

The resulting sounds have been described as weak knocks, thumps, bumps, and booms. Like the snappers, a great number of other fish use the swim bladder and associated muscles to produce sound. Included in this group of vocal fish are sea robins, some wrasses, damselfish, angelfish, drums, croakers, and many of the sea basses.

Sea Lions

Visitors to certain seacoasts may hear the barking sounds of seals or sea lions in the off-shore waters. Most of us have heard these bright and attractive animals vocalizing at zoos. Seals, sea lions, and walruses have well developed vocal cords and make use of them to bark or growl for a variety of reasons. In addition, they make clicking sounds underwater, which also originate in the larynx or just behind it. When sea lions make these clicks, they may have their mouths and nostrils closed or partially opened with a consequent exhalation of a trail of bubbles. Walruses click their teeth but also produce bell sounds using air bags they have in their throats.

A major reason sounds are produced is for recognition, a very important function between a mother sea lion and her pup. Pups produce a sound described as "aa, aa, aa." This provides a means of recognition from a distance for the mother sea lion. Scent prob-

ably plays the most important role in recognition up close. Clicking sounds may be part of an echolocating system. Other uses for vocalizing include threatening intruders, gaining attention, attracting a mate during courtship, and asserting dominance. Sea lions and other pinnipeds have a distinct social order in which one animal is dominant over all others in its group and others have ranks below the dominant one. If one animal challenges the dominant animal of its group, vocalizations are one of the ways in which

A sea lion pup rests quietly at its mother's side after it has finished nursing. Vocal communication between mother and pup keeps them from being separated from each other.

the number one animal asserts its leadership. This usually takes the sound of "goh, goh, goh." Other adult sea lions emit sounds more like "ga, ga, ga." Pups start out with a high-pitched sound, which deepens with age and growth. Males have deeper, more resonant voices and are considerably larger and heavier than the females.

Whales

Some species of whales have been so reduced in numbers there may be too few of them left for their species to survive. A major problem for a sexually mature whale is finding a mate in the millions of cubic miles of ocean.

Researchers have recently turned up some interesting theories based on data accumulated from underwater listening and acoustical studies. Sound travels fastest in warm water and in dense water. In the oceans a deep sound channel exists between the warm surface layer and the dense layer below 4000 feet. When a whale's loud calls are directed obliquely, they travel in waves in this channel. The calls are thus reflected between the warm layer and the dense layer with very little loss of energy. The sound waves travel up and down in this channel at a regular rate for great distances, nearing the surface every 35 miles.

Natural background noises in the sea limit the distance that whale sounds can penetrate and be distinguished. These back-

ground noises include the sounds of the myriads of shrimp, the hydrodynamic sounds of fish and other animals passing through the water. A major background sound that blocks out whale calls is the noise of propeller-driven ships.

When a rorqual whale calls as it cruises in the surface waters searching a mate, its voice carries for about 50 miles. If the sound of all the ships at sea were stilled, the rorqual's voice could be heard by others of its species for close to 150 miles. If all other background noises could be eliminated, its voice might be heard for a few hundred miles. In reality, though, there will always be background noises. But if the whale were in the deep sound channel under optimum conditions, it could theoretically be heard a few thousand miles away. Low-frequency sound travels farthest in the sea and is mainly limited to the power of generation. In the case of some rorquals there is a relatively great power output and the frequencies range as low as 20 hertz (blue whale), making relatively long distance communication at least a possibility.

Chapter V. Knowledge Through Scent

Blood gushes from a wounded fish and water carries it down current. Animals in the path of the blood sense it in different ways. In the distance some fish pick up traces of the blood with their olfactory receptors. They are said to smell it. Those close up may move in and mouth the wounded, bleeding fish and thereby sense directly the flood of the blood molecules. These animals are said to taste it.

All animals are sensitive in varying degrees to changes in the chemistry of their environ-

> "The sense of smell provides
> for reception of chemical
> stimuli from a distant source."

ment. A change may be caused by the release of blood from the wounded fish, by release of "alarm substance" from a frightened animal, by man's wastes dumped into the sea, by changing salinity, and by other factors. Some animals secrete substances (pheromones) that convey information to others and often evoke specific responses.

The olfactory sense, smell, provides for the reception of chemical stimuli from a distant source. It is much more sensitive than the taste sense, which relies on physical contact. The olfactory sense can respond to a few molecules, but the taste organs must contact large concentrations of molecules.

The senses of taste and smell are developed to different degrees among invertebrate animals. In the most primitive animals, the protozoans and the sponges, these senses exist, but not in the same manner we usually define them. In animals a little higher up on the evolutionary scale, the senses of taste and smell exist and function to some degree. Molluscs have sensory patches (osphradia)

located close to their gills, which react to chemical changes—that is, they are able to smell. Some also have taste buds. Some crustaceans have chemosensitive cells over their whole bodies; these cells are located directly beneath the animal's shell and are connected to the surface by minute pores and ducts. In fish the sense of smell lies in the nasal pits. These pits are lined with tissue that picks up scents and relays impulses to the olfactory lobes of the brain. In some species these lobes are enormous in comparison to the rest of the brain; in others they are a minor part of the brain. In some fish the nasal pits are connected to the mouth, but in most they are not. In no fish, however, are the nasal pits connected to the respiratory tract as they are in terrestrial mammals.

Despite all the highly specialized sensory equipment that marine mammals have, none has a sense of smell that operates underwater. The marine mammals are air breathers, and keep their nostrils or blowholes closed while submerged. And what is known of the sensory ability of the sea mammals indicates that they have, at best, a poor olfactory sense in air. Similarly, little is known of their sense of taste, but indications are that most have some ability to taste.

So with the exception of the marine mammals, all the animals in the sea have a sense of smell. With this sense they can discriminate between individual animals, recognize the sex of others, be alerted to the presence of enemies, friends, and their young, notice the presence of food and track it down.

Sea lions often recognize others around them by sound and smell. These senses protect a newcomer from unnecessary aggression and warn the group of unwanted trespassers. Here a sea lion puts its nose to a diver's finger.

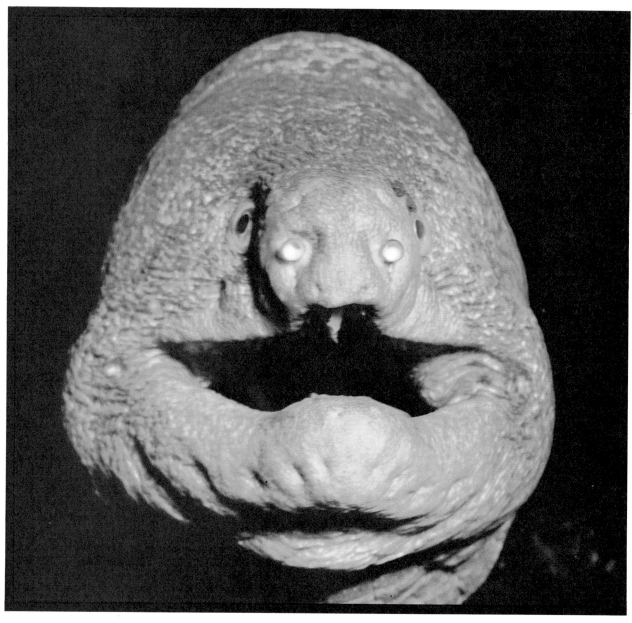

Green moray eel. *A well-developed sense of smell guides this nocturnal hunter to its prey. Note the prominent tubular nostrils.*

Moray Eels

With many fish, water for breathing, or smelling, circulates as they swim. But moray eels, just as other sedentary fish, must circulate water by other means.

The moray's two nostrils differ from those of many other fish. Fleshy tubular extensions reach from their anterior openings. Inside the nostrils, beating hairlike projections propel water through the nasal cavities. The moray feeds at night, preying on other reef dwellers and scavenging in the vicinity of its home. With no sunlight to see by and a reduced lateral line to "feel" by, the moray is informed of the presense of food and is guided largely by its sense of smell.

*An **octopus** rolls its many arms back, exposing the suckers on their undersides. The suckers are tasting and smelling organs.*

Octopods

The whole surface of an octopus's body is sensitive to waterborne chemicals. The two rows of suckers that stud each arm contain receptors, which can sense by smell, taste, and touch, and are very precise.

Experiments conducted with blind octopods have shown that the animals depend more heavily on tasting and smelling than on feeling to learn about their surroundings. For example, an octopus, which is given a choice between an empty crab shell (the right feel) and a porous stone impregnated with animal juices (the right smell and taste), will pick up the crab shell and discard it, but pick up the inedible stone and chew it.

Keyhole Limpet and Sea Star

The keyhole limpet is a small animal usually found holding on to rocks with its fleshy foot. Its normal position is seen above; its siphon is beneath the level of the keyhole, and its flexible mantle is beneath its body, protruding just a bit from the sides. Since it cannot move easily, and since it has no eyes, ears, or nose in the normal sense, one wonders how it can get information about danger and how it can do anything about it once it has received the information. Apparently we need not worry about the keyhole limpet, for experiments have shown that it has a keen chemical sense and an excellent apparatus for self-protection.

Through its sense of smell the keyhole limpet can sense the direct or indirect presence of several species of sea stars from quite a dis-

Keyhole limpet. Because of the hole at the peak of its shell, the keyhole limpet (above) is also called the volcano limpet. If it feels no threat, it moves along the bottom with its shell uncovered.

Sensing danger. At the approach of a California sun star (opposite page), the limpet covers its shell with its fleshy mantle, rendering itself safe from predation by the sun star.

tance. The limpet then makes some remarkable changes. It does not move away, rather its siphon and mantle move to cover and protect it where it stands. The siphon comes up out of the keyhole and swells in such a way as to leave no opening for the sea star. At the same time, the mantle fold stretches and moves up the sides of the limpet, stopping when it reaches the top, where the siphon has protruded from the keyhole. The limpet is then capable of elongating the foot with which it is attached to the rock. The "help-

less" little animal has constructed a wall around itself and is safe from the sea star. Since the "wall" is constructed with the fleshy material of the mantle, the sea star's tube feet, with which it holds on to its prey, are useless. They simply cannot hold on and, in fact, seem repelled by the mantle.

Without its olfactory sense, the keyhole limpet might never react in time. In fact, experiments were conducted in which limpets were placed in water that *had* held sea stars,

and it took the limpets only a few minutes to sense that their enemies had been there.

The many-legged California sun star is one of the animals whose presence provokes this protective response in the limpet. Although the sun star inhabits the same locale as the limpet and will eat them, limpets do not make up a large part of the sun star's diet. Perhaps without the mantle response to the chemical produced by the sea stars, limpets might be heavily preyed upon by the stars.

Cone Snail

The deadly cone snail often hides itself in sand on the sea floor with only its siphon extended upward into the water. Through the siphon, water is pumped to the gill, which in turn extracts life-sustaining oxygen. Adjacent to the gill is a patch of sensory tissue, the osphradium, which is the cone snail's receptor for odors. Using the osphradium, cone snails can sense distant prey and track it down until they are close enough to touch it. Then they can bring into play their impressive venomous weaponry.

On the opposite page the **cone snail's** *siphon is extended to pump water for respiration. The water also brings chemical messages.*

At the right (inset) we see the cone snail after it has **emerged** *from beneath the sand. Having detected the presence of a meal, it goes after it with siphon still extended and working.*

Murex Snail

In the photograph above, two murex snails are preparing to feast on a live clam. Although many snails are scavengers, the murex snail clearly prefers the meat of a live clam or oyster to a dead one. And it has developed an elaborate apparatus for getting at the meat of its victims.

How can the snail tell the difference between live and dead clams? Apparently the answer is chemical. Experiments have shown that small amounts of a chemical substance emitted by live oysters is enough to attract the murex and other related snails to their prey. This sense of "smell" is very well developed, and although the exact distance at which it operates is not known, it is believed to be considerable. Scientists have been unable to

Preparing to feast. Attracted to a live clam, two murex snails (above) will devour it. These snails sense chemicals in seawater.

Alert scallop. Rows of bright eyes of the scallop (right) are augmented by many tentacles which are sensitive to both touch and chemicals in the water.

discover from what part of the oyster the fluid comes, but that is not really important —what does matter is the snail's sensitive receiving apparatus for it.

When the snail discovers this scent in the vicinity, it goes in search of the meal, finds it, and attaches itself to its prey. It then bores a hole in the shell by means of a chemical and a filelike radula, then rasps out the victim's tissue. A close relative of the murex snail is the notorious oyster drill that attacks commercial oyster beds and causes losses that run to millions of dollars annually.

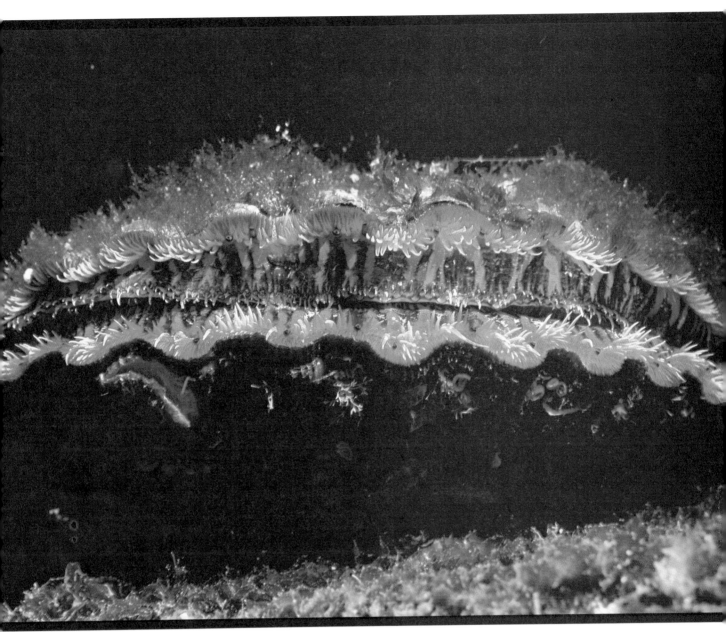

Scallop

The scallop, especially when compared with the other lower animals, is highly sensitive and alert. It has three distinct methods of receiving messages. Tentacles extending from its mantle are sensitive to touch and allow the scallop to know where it is going and whether what it is touching is friend, foe, a meal, or a place to rest. At the root of these tentacles, rows of many brightly shining eyes line both the upper and lower rims in the shell. Although these eyes are primitive, and only allow the scallop to distinguish between light and dark, the information they give is helpful and tells the scallop about its surroundings. The scallop's tentacles are also sensitive to chemicals in the water, and the shellfish uses this highly developed sense to protect itself. Its chemical sense helps alert the scallop to an approaching octopus, sea star, or other predator. And very often the scallop can propel itself away using a jerky form of jet propulsion.

*The **American lobster** has an acute sense of smell to pick up chemical communications.*

Atlantic Lobster

The Atlantic lobster has been found to have a chemical language. The chemicals used in this language are called pheromones. These carry messages of different sorts. For each message to be sent, there is apparently a different pheromone. There is one identifying the sex of an individual lobster. Another

may tell of aggressive intent or intent to defend territory. Others tell of desire to mate and can be detected at a distance. They are apparently strong and persistent.

When researchers placed a sex pheromone from a just-molted female lobster in with an aggressive male lobster, the male began a mating dance although there was no other lobster in his tank.

Sea urchins massed on brown algae, their prickly spines alert to chemical changes in the water.

Sea Urchins

Sea urchins are a bundle of nerves. Every spine, every tube foot, and every pedicellaria or pincer is a sensor. The nerves, which are found in all these parts, sense chemicals in the water around them. In addition, sea urchins have another rich supply of nerves in the jaws of the pedicellaria and just under their outer surfaces which are sensitive to touch.

When a sea urchin is ready to spawn, it must wait until its chemoreceptors advise that another sea urchin, which is also ready to spawn, is in the vicinity and that it is of the opposite sex. When the two sea urchins sense each other, spawning can begin with some hope of success.

85

Sharks

More than 200 species of sharks have been identified. Some are bottom dwellers; usually they eat smaller bottom-dwelling fish, crustaceans, and molluscs. A few species are deep-water fish, seldom seen except when captured by probing scientists or photographed by abyssal cameras and by bathyscaphes. A number of other species are pelagic sharks, ranging through the open ocean and occasionally cruising in close to shore. These include the requiem shark family, often considered to be dangerous to man but not consistently so, as many divers know.

Sharks are well equipped to find food even in the wide open spaces of the ocean. They use vision in tracking down their prey, but their main clues in finding distant food are pressure waves and odor. Precise, short-

Gray reef shark. This large, well-fed shark effectively uses sound, sight, and smell to locate food. It rarely ventures in the open oceans.

Blue sharks. To the right, a pack of slender blues has been attracted by the smell of food. Scientists baited the water in order to study shark behavior.

range guidance is provided by vision. Medium-range information by smell. But if their prey is a thrashing, already injured fish, they may pick up the desperate beat of its tail at great distances and suddenly appear from nowhere.

Sharks are not hungry very often, but when they are, they use all their senses at maximum acuity to help them search for food. An underwater explosion, splashes from exuberant swimmers, stones thrown into the sea, and struggling fish sounds played back in the ocean have all been known to attract sharks from beyond their range of vision.

Strictly speaking most sharks are not school-ing fish. There are occasions, however, when they gather in numbers, swimming around and near each other in many differ-ent directions. Such is the case when fisher-men seeking to catch sharks bait the waters with chum (ground up bits of fish and fish oil). Even if there are no sharks in the area when fishermen spread the chum in the waters, there will soon be many. News of the free food spreads as fast as the currents carry the bits of fish and droplets of fish oil through the sea. Soon there may be scores of them swimming through the area. That's when the fishermen start throwing food to them—on hooks at the end of their fishing lines. Thrashing sharks, which have been hooked and are fighting to free themselves, occa-sionally become food for the other sharks. Their blood attracts even more sharks to the area. To date, all attempts to develop a shark repellent based on smell have been failures, but recent experiments using repel-lents *developed by fish* are promising.

Chapter VI. Electricity and Communication

Fish are the only animals that are able to produce and deliver a considerable electrical charge. However, all animals, terrestrial as well as aquatic, within their specialized muscle, glandular, and nervous tissue produce minute electric charges, once called "animal electricity." A few families of fish have developed this ability to the point where they have control over it and are able to make use of the information it provides.

Most of the fish producing electrical charges live in fresh water, but in the sea a few primitive cartilaginous fish and one bony fish produce electricity. Electric organs are made of highly modified and specialized muscle tissue no longer used for movement.

Producing electricity must have been beneficial to the ancestors of today's living batteries, and the process of natural selection passed it on. The ability is not equally developed in all current-producing fish. Some produce a charge with very high voltage but low amperage; others produce a burst of electricity with relatively low voltage but very high amperage; and still others discharge a current with both low voltage and low amperage. The probable reason for these variations is the use to which the electricity is put. Those with the most powerful shock use it as an offensive and defensive weapon, discharging when they detect prey or when they are disturbed. Other animals, with less powerful abilities, use their low power discharges to supplement information received by their other senses. The electric field they generate aids them in navigating and finding food and may help them locate mates.

The tissue that makes up the batteries of electric fish looks something like a stack of coins, each coin being a single cell called an electroplaque. An electroplaque resembles a cream-filled cookie with a different type of wafer on either side of the filling. The two cookie wafers represent the different membranes on each side of the electroplaque's jellylike center. Nerve fibers connect to only one of the membranes.

Each electroplaque is a tiny battery, producing a small charge between the two membranes. All the electroplaques in a single column face in the same direction, so they

> "Electric organs are made of some highly modified and specialized muscle tissue no longer used for movement."

are connected in "series." Batteries connected in series add their voltage outputs to one another, so the column of electroplaques creates a difference of potential that is the sum of the many small voltage differences of the individual electroplaques. The charges are released on electrical impulses from the nerve fibers.

In some fish, such as the electric eel, there are only a few very long columns of electroplaques. These run lengthwise, parallel to the fish's backbone. In others, like the torpedo ray, there are many more columns, but these are not nearly as long, and they are situated perpendicular to the backbone. The longer columns produce a high voltage charge, and more numerous columns produce a charge with high amperage.

Olivella snails. These little burrowing animals travel beneath the sand, leaving long trails behind them. Tests on other species of mud snails show that their direction can be influenced by the intensity of a magnetic field.

Torpedoes

Among the flat, cartilaginous fishes that we call skates and rays there are several species that have electric organs, which can generate significant amounts of electricity. These electrical charges serve an important function in these fishes' lives. What that function is varies with the species and with the amount of electricity that is generated. The species that produce these electrical charges include the torpedo ray, the narcine ray, and the thornback skate among saltwater fish.

Both the torpedo and narcine rays have a pair of electric organs on each side of their heads. The larger of these paired organs can generate enough power to stun a would-be attacker or the fish's prey. One species of torpedo ray has an additional pair of electric organs. It is speculated that these are used as a navigational aid. As they move about their environment, they may carry their

electrical fields with them. When that field is disrupted by an obstacle, the ray may sense the disruption or distortion of the field.

The skate's electric organs are more of a mystery. It has a pair of electricity-producing organs which are formed of modified muscles, one on each side of the base of its tail. No one knows how or why the skate uses its ability to generate electricity. The skate is reluctant to discharge current and does so only after considerable prodding.

*A **torpedo ray** rests on a sandy sea floor ready to ward off any intruders with an electric charge.*

The skate's electrical charge is of low output, only about four volts, while the torpedo and the ray may discharge 40 volts or more. One torpedo was measured as discharging 220 volts and a fairly high amperage, but this is the highest voltage observed in the torpedo. (Voltage is the amount of force with which the electrical current is driven; amperage is the strength of that current.)

The Ampullae of Lorenzini

Sharks, rays, skates, and chimeras, known collectively as elasmobranchs, are among the most primitive of fishes, with skeletons of cartilage instead of bone. But in their primitive state they have some advantages over the more advanced and more fully developed bony fishes. One of these advantages is a special organ called the ampullae of Lorenzini. These animals have many ampullae scattered over the top and sides of their heads—that are evidenced outwardly only by small pores in the skin.

The ampullae of Lorenzini are small sacs filled with a jellylike substance. These sacs are connected to the surface skin by ducts leading to pores. The linings of the sacs are sensitive tissue with many folds, bumps, and other irregularities. The ampullae are able to sense electrical fields from various sources in the water. Careful measurements showed that the ampullae can sense as little as one-tenth of one microvolt (one one-millionth of a volt) of electricity.

Knowing that electrical discharges sensed by an elasmobranch caused its heart to slow, one marine biologist set out to determine if the ampullae help the elasmobranchs in hunting. To do so, he first measured the electrical impulses generated by various movements of a common flatfish, the plaice. He found that in moving its gills in normal respiration, the plaice generated about 1000 microvolts. He also determined the electrical charge generated by the plaice during burrowing and found it was higher than that produced during normal breathing. He put the plaice in one tray and a thornback ray in another tray and attached to the ray an electrocardiograph machine to measure its heart action. The two trays were connected only electrically. The ray could not sense the presence of the plaice in the other tray except electrically. When the plaice burrowed, the researcher noted the ray's heartbeat slowed. When it stopped, the plaice's heart speeded up again. When the plaice breathed by moving its gills, the ray's heartbeat slowed again. Since it is already known that nervous impulses controlling these movements are electrical, the researcher was able to verify the ray's ability to sense the electricity generated by the plaice's movements.

In further experiments, the same Dutch researcher trained a group of sharks and rays to eat in an area directly over a pair of electrodes buried in the sand bottom. When he fed the fish, the current was turned on and emitted four-tenths of a microvolt. Then he stopped putting food out, but when he

This leopard shark, one common in the shallow waters of California, is able to perceive the electrical field of other living animals.

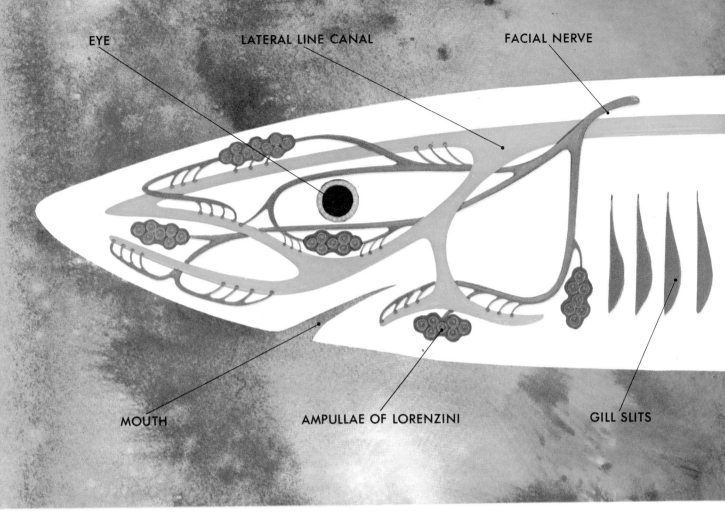

EYE   LATERAL LINE CANAL   FACIAL NERVE

MOUTH   AMPULLAE OF LORENZINI   GILL SLITS

Ampullae of Lorenzini. Details of the anatomy of the ampullae of Lorenzini are shown in this diagram. Ampullae have been reported to sense electricity and possibly temperature and pressure.

turned on the tiny electrical charge, the sharks and rays swarmed about the electrodes, even uncovering them and snapping at them. Thus, he showed again they could sense minute charges of electricity and could also trace it to its source. These tiny charges could only be sensed at close range so he concluded the fish use the information provided by their ampullae of Lorenzini for close-range hunting. This close-range sensing is especially valuable to the fish as they approach to bite. Because the placement of their eyes keeps them from seeing immediately in front of them, they lose sight of their prey when they approach close enough to bite, and they could miss capturing it.

The ampullae have been thought to have other functions than sensing electric fields. One possibility has been described as a sense of water depth (hydrostatic pressure). The ampullae have also been experimentally

shown to respond to changes in salinity and temperature, but have never been definitely confirmed to fulfill such a function in the intact animal freely swimming in the sea.

When tissue from the lining of the ampullae was taken from a fish and studied, it was found to emit rhythmic pulses of electrical energy in minute amounts. These rhythmic pulses varied with the stimulation given the tissue. When the temperature was lowered, pulses speeded. When temperature was raised, pulses slowed. The tissue also reacted to slight changes in salinity, leading some researchers to believe the ampullae can sense variations in salinity.

93

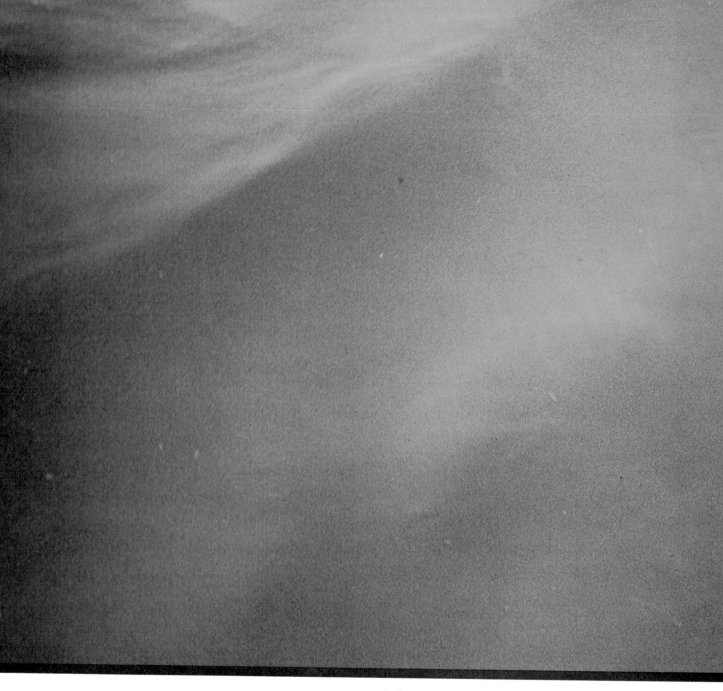

Open-Ocean Shark

When a coil of wire is rotated within the field of attraction of a magnet, an electric current and resultant field is generated. Electrical charges are generated in the oceans of the world in much the same manner. Great rivers, the ocean's currents, flow across the face of our planet, and in doing so pass through the magnetic field of the earth. They become the equivalent of the coil of wire responding to the magnet. Electrical fields are generated. We know that sharks have special organs that can sense electrical stimuli, including that generated by the ocean currents. But do sharks use the electrical field in the oceans to navigate? If a mechanism for navigating by electrical stimuli does exist in sharks and can be discovered, it might help us devise underwater navigational systems that are improvements over our present ones.

A clue to such a mechanism may be found through a physiological study of the shark's

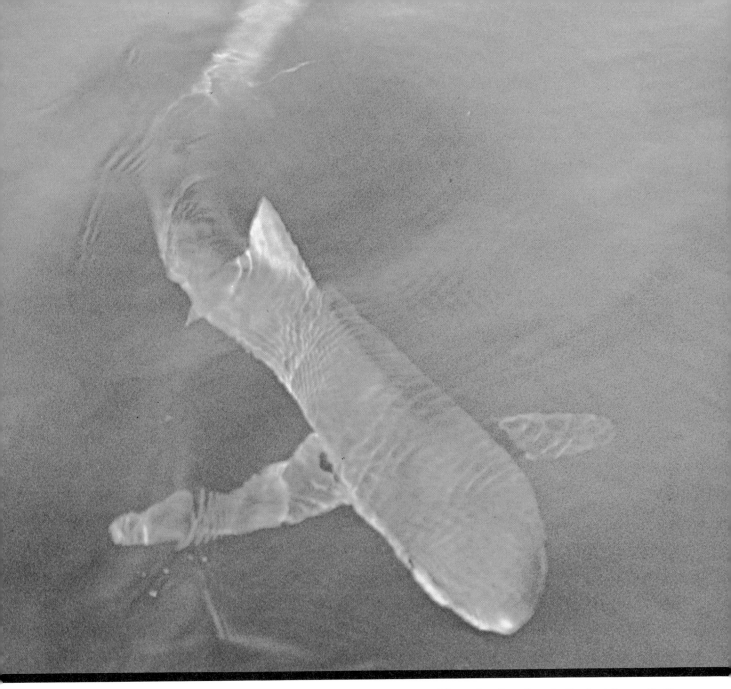

brain and its response to various stimuli. This may lead to greater knowledge of how sharks use their electrical sensing ability to locate food. It is already well documented that the enlargement of certain parts of the brains of some fish relates to their method of finding food. Perhaps in a similar way the physiology of the shark's brain can tell us if and how it uses these electrical impulses in navigation. Answers might be found through experiments with the living animal by giving it problems in navigation to solve. Such a

*A great **blue shark** swims lazily just beneath the ocean's surface. Primitive sharks may be able to teach us something about oceanic navigation.*

problem might force the shark swimming in a certain direction as a response to an electrical field to change direction when that field was reversed.

Sharks, although often called primitive creatures, have continued to evolve well-developed systems like their electrical sensors over many millions of years.

Stargazer

Lying on the ocean bottom as they usually do, concealed by the sand around them, stargazers are constantly looking skyward (or so the person thought who first named this weird-looking fish). And its scientific name, *Astroscopus,* which means "stargazer," assures that it will continue to be so called. But the name comes from the physical construction of the animal. Its eyes are located on top of its head and are directed upward—not, however, at the sky but at the waters directly above the stargazer where its prey is most likely to pass. Some members of the stargazer family have a small wormlike organ attached inside the mouth. When another fish tries to bite it, the stargazer shuts its mouth and has a meal.

The stargazer has another tool besides camouflage and trickery. It has electricity-producing organs in a small patch of muscle behind each eye. It is one of the few marine fish species with the ability to produce sig-

nificant amounts of electricity. Only some members of the skates and rays share this ability with the stargazer. No one knows how the stargazer uses its electrical potential. The electric organs of the stargazer are much smaller than those of all the other electricity-producing fishes. But like these fish, the stargazer has specialized muscle fibers, each fiber a separate electroplate.

In some fish the electric charge is used to stun and capture live food. Some use the electricity to fend off attackers. Others use

Eyes up. The electricity-generating stargazer usually lies almost buried in the sand with only its eyes and mouth exposed; occasionally it swims awkwardly to a new hunting location.

it in navigation. When it is used to navigate, the fish, whichever species it is, creates an electric field that surrounds it. An object coming nearby may produce an irregularity within the field. The fish knows by sensing such an irregularity that there is an obstacle or interloper in its vicinity.

Chapter VII. Sounds and Pressure Waves

When a solid object moves through air or water, it causes the molecules around it to move. The molecules in front of it, are pushed together, or compressed. This compression of the molecules is called a pressure wave. The molecules behind the object spread out to fill in the void left by the moving object.

Under some conditions we can feel a pressure wave on our skin, as we do when we are standing near a road and a car or truck whizzes by. Then we are buffeted by the wave of air being pushed ahead of the vehicle. When pressure waves have certain other characteristics, we can no longer feel them on our skin, but sense them with our ears as sounds, which are a form of pressure waves.

If the object does not move continuously in one direction but moves back and forth about a fixed point, it sets up pressure waves first on one side (as it moves one way) and then on the other (as it moves in the opposite direction). If the object moves back and forth again and again, it creates a series of pressure waves on each side, which travel away from the object. Each complete movement is called a cycle.

If we count the number of cycles that occur in a period of time, we get the frequency of the movement. When discussing sound, the frequency is timed in seconds. When the frequency falls between 20 cycles per second (hertz [Hz]) and 20,000 Hz, humans can hear them. Frequencies in this range are called "sonic." If the frequency is less than 20 Hz, it is called infrasonic, and if it exceeds 20,000 Hz, it is termed ultrasonic.

Whales and dolphins have highly developed hearing, particularly sensitive to ultrasonic sounds. Hearing is probably their most important sense, but their ability to hear well was turned against them by early Japanese whalers. These men drove dolphins and whales into shallow bays by beating on the sides of their boats with hammers.

Most fish can best hear sounds that are in the 200 to 600 Hz range. Some species of fish are sensitive to sounds with frequencies as low as 10 Hz, and others may hear frequencies over 10,000 Hz. It is difficult to determine whether a fish hears or "feels" the pressure waves with its lateral line because sound pulses or turbulences are all basically the same phenomenon.

The lateral line system is especially sensitive to low-frequency vibrations caused by movement underwater. It gives fish a sort of long distance sense of touch and keeps it informed about nearby animals and objects. This is especially helpful when the fish is swimming in dark or turbid waters.

Pressure waves may also play an important role in the social behavior of a fish. In the ritualized fighting before mating, some fish fan their opponents with their tails to establish dominance. Some other fish fan their mates as part of the courtship procedure between male and female. The vibrations set up by the fanning probably convey appropriate messages to the animals involved.

*The **blind cavefish** lives in a lightless, subterranean world, and over millennia it has lost its ability to see. Its other senses therefore become even more important to its survival. Strangely, the cavefish does not have the lateral line system common to most fish for detection of pressure waves. Instead, it has numerous sensory papillae on its head to sense vibrations caused by nearby movement. It apparently has difficulty sensing stationary objects, since in aquariums it continually bumps into objects as well as the walls of its tank.*

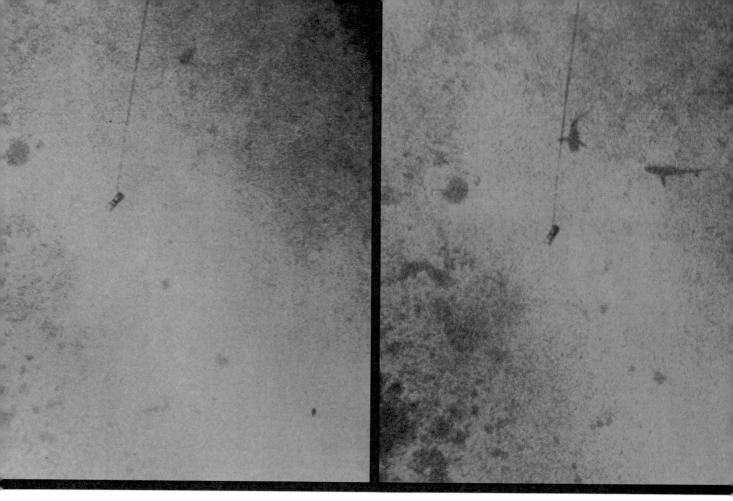

Turning on the sound transducer. When an intermittently transmitted pulsing sound was first turned on, no sharks were in the area.

Ten seconds after the transmitter was turned on, the first few sharks began to arrive at the sound source.

Sharks Attracted to Sound

Sharks depend on their sense of hearing to some degree and are especially attracted to low-frequency pulsed sounds that are intermittent and irregular in character. These are similar to struggling sounds of fishes in trouble. A number of experiments demonstrate that sharks are drawn to the source of such sounds transmitted underwater. And because some of the sharks continue arriving after the sound is shut off, some believe the sharks may have the ability to remember direction.

Two researchers working at Eniwetok atoll in the western Pacific ocean conducted a series of acoustical experiments with sharks. Dr. Donald R. Nelson and Richard H. Johnson ran their tests at a number of points along 10 miles of coral reef waters about 65 feet deep. They used taped, electronically produced sounds designed to simulate fish struggling. Five species of sharks were attracted to the sound emissions. Of the total number, more than 80 percent were gray reef sharks. To make sure that the sharks were actually attracted to the area by sound, the researchers lowered the transducer into the water without turning on the sound and observed for five minutes. Then they turned on one of a series of three sounds for five minutes to see the results.

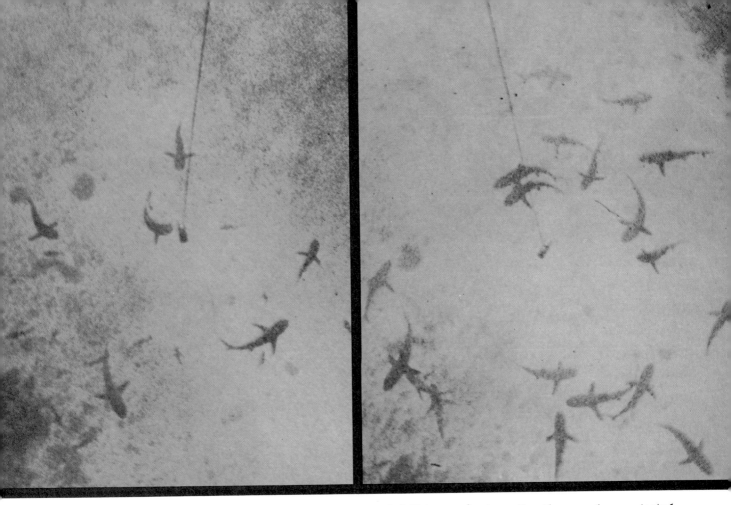

Half a minute after the sound started, several more sharks had arrived and were circling the transducer. One shown here is about to bite.

*A full **two minutes** after the sound was started,* many sharks circled the transducer. These scenes were enlarged from 16-mm motion-picture frames.

Testing at the various stations showed that a total of only 44 sharks were seen in the areas when no sound was being produced. When the sound was turned on and pulsed signals emitted, a total of 253 sharks were drawn to the sound source.

The three different broadcasts transmitted at various times were designed to determine if sharks were more responsive to particular patterns of sound. The researchers used a low tone (25 to 100 Hz), much lower than a man's voice. The first experimental "train" of sounds was transmitted at 10 pulses per second continuously. That is, the tone reached a peak and subsided 10 times a second. They sent the same pulsed tone intermittently as the second pattern. The third pattern was the same tone but pulsing intermittently at a varied 7.5 to 15 pulses per second. The first sound pattern drew 51 sharks; the second brought 102; and the third brought 100. As a control the same areas were observed just before sound was turned on. Twelve, 18, and 14 sharks were seen in the silent control periods.

The sharks remained active for some minutes after the sound was turned off. They investigated the researchers' boat and observation float. The men were unable to enter the water because of the excited sharks.

TV camera setup. *A researcher (above) cleans the housing that encloses the TV camera.*

Sound projector. *A goatfish (below) cruises the artificial reef that houses the sound projector.*

Underwater TV for Research

Scientists are now using underwater television cameras as constant spies in connection with underwater sound projectors and hydrophones to study the effects of sound on fish behavior. By staying away from the area under study and observing on a TV monitor at shoreside, the diver-scientist does not disturb the sea life that may gather at the site, and the animals therefore react as they would in the absence of man.

Such an underwater TV set is routinely used by the divers of *Calypso*, either on a tripod on the sea bottom or attached under the keel of a small boat. A TV camera bolted under a motor launch and aimed toward the rear made it possible to film the way sharks bite on a bait towed at six knots. In the Bahamas, researchers from the University of Miami and the American Museum of Natural History were able to observe on a shore TV monitor the effects of various sounds on free-ranging reef fishes and sharks.

Fish attracted to reef. *Joining the goatfish are a yellowhead wrasse and a damselfish (above).*

TV monitor screen. *In the laboratory the TV screen (below) shows a snapper swimming past.*

On TV screens it is possible to study reactions to various sounds of fish and mammals when they are hearing playbacks of either their own sounds or the sounds of those marine creatures they are most concerned with, either as prey or as predator.

It was not known until about 1903 that fish could hear. Since then many experiments have taken place. The more that scientists discover about how fish react to various sounds and noises, the more they will be able to help fishermen attract fish or keep unwanted species away. Tuna fishermen may someday be able to keep dolphins away from their nets by playing to them the attack calls of killer whales. Perhaps fish someday can be induced to breed in certain areas by playing back to them the sounds that are usually associated with the breeding season. Sounds can also be used to draw migratory fish away from danger.

All this research may pave the way to huge, efficient sea-farming enterprises to replace today's fishing industry.

Vocal Damselfish

Small pugnacious damselfish are common to many Caribbean coral reefs. A number of them are highly vocal and use sound to communicate. Recently some of these fish were subjected to an interesting investigation made in the waters off Bimini Island. Scientists studying the acoustic behavior of fish have noted that one species, the bicolor damselfish, uses sounds in courtship and aggressive behavior. When previously recorded "chirp" and "grunt" sounds were played in the water, males engaged in a courtship dance. The courting male damselfish has a peculiar "dip" pattern of swimming which was always associated with the "chirp" sound. These sounds induced reproductive behavior in fish even at times of the year during which the fish are sexually inactive. Aggressive behavior was associated with "pop" sounds in both laboratory and open sea experiments. Played in the water, this sound inhibited all reproductive behavior.

*A handsome **blue chromis,** member of the damselfish family, moves warily over a coral reef.*

Barracuda

*A great **barracuda** moves slowly through the shallow waters of a coral reef.*

Like a sleek and silvery submarine, the great barracuda cruises, constantly alert, halting occasionally to watch for prey or to size up its neighbors. Although the barracuda uses its vision in its sometimes violent search for food, it also is constantly sensing pressure waves of water reverberating from its surroundings.

Using their ability to sense changes in pressure waves, barracudas locate far-off schools of smaller fish. In turbid waters they also depend heavily on their lateral line system. Bodies only partly submerged and thus not readily recognizable by the barracuda may be subject to attack. In the few reported cases of attacks on humans, the water was shallow and turbid, and the barracuda mistook a moving hand, foot, or leg for a potential meal. The barracuda, unseen, may have been mistaken for a small shark.

A Mediterranean sea robin, with the sensory rays on its pectoral fins folded back, swims just above the sea floor.

Sea Robin

As they do for most fish, the pectoral fins of sea robins and armored sea robins serve as organs of propulsion and stability; but the three lowest fin rays are free from the rest of the fins. Nerve endings in these unattached spiny rays can perceive molecules of matter drifting by in the water, and thus the fish can smell or taste with them.

In addition, the armored sea robins have a pair of barbels on their chins, which can taste or smell. By dragging the chin barbels along as they stir up the bottom, the armored sea robins can taste what is edible in the material on the bottom.

Fiddler crabs. Their eyes on the ends of tall stalks, two fiddler crabs wave their large claws, possibly warning each other not to approach any closer.

Fiddler Crabs

The fiddler crab lives in dense colonies on beaches in many tropical and temperate areas throughout the world.

Fiddlers spend much of their time near the entrances to their burrows. They are able to hear vibrations through the ground they stand on. Males produce either a rapping sound or a honking sound as courtship signals. The rapping sounds are produced by drumming on the ground with the large claw. Leg movements are thought to produce the honking sound. The females can hear the low-frequency sounds the males produce from more than 30 feet away.

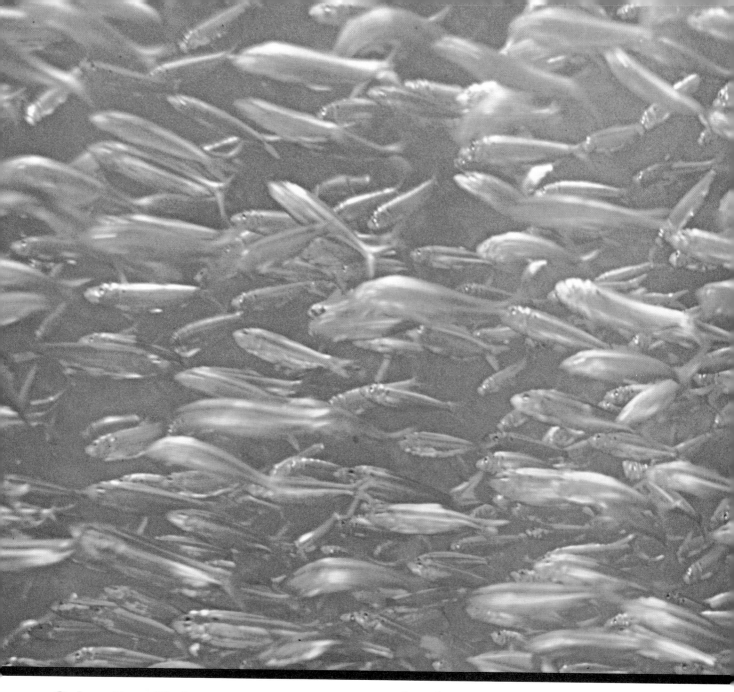

Schooling Fish

Of more than 20,000 species of fish recognized by taxonomists today, some 4000 are known to school. That is, they move together in more or less evenly spaced parallel orientation and more or less tightly packed groups, all acting in a similar manner with no apparent leader.

To try to understand the mechanism that enables fish to swim close to each other with-

out interference, we must remember that water is nearly incompressible and somewhat transparent. There seem to be two principal ways in which fish are able to maintain close order formations as they move in the sea. One is by visual contact. The importance of vision in what is called "prolonged maintenance of parallel orientation" has been recognized for many years. But schooling fish have been equipped with blinds and still continued schooling, though with variations in pattern. Yet there is still another fac-

tor more important than sight in the case of fish schooling in turbid waters—that is the lateral line system discussed on page 39. In experiments fish were alternately deprived of the opportunity to use their lateral line and their sight to orient themselves to each other in a tightly packed school. When they were permitted to use their lateral line systems, the fish were able to school effectively. In areas of poor visibility, the lateral line enables the fish to maintain a close, but appropriate, distance among themselves. If

*Almost like **sardines** in a can—except they're alive —these sardines move in unison through the sea.*

they swam too close to each other, the vortices of turbulence caused by their swimming motions and their passage through the water would disturb the neighboring fish in the school. This would ultimately result in the school being strung out in long lines, with individual fish getting farther apart. In clear water, the school expands, and sight becomes more important to the schooling fish.

Challenging Bellows

More than 100 miles to seaward from Baja California lies Guadalupe Island. Here the northern elephant seal population gathers annually to stage its mating ritual. The bulls immediately begin establishing the social order. The males confront one another, bellowing loudly, and sometimes fighting violently. Any male can be challenged at any time by any other male.

The dominant bulls must defend themselves against the challenger's onslaught, since if they lose a battle to a subordinate, they drop down several positions. Most of the aggression does not get past the threat stage, when loud vocalizations and chest-to-chest standoffs occur. The bulls trumpet through their long noses, which serve as resonators to intensify the bellowings. When one male approaches another to challenge him, he trumpets from three to fifteen low-pitched, gut-

teral sounds to signal his intentions. The threatened male then lifts his mountainous three-ton hulk to meet the challenger. The two bulls trumpet back and forth. If neither gives ground, they resort to combat, slashing at each other's throats.

To test the effect of vocalization on resting bulls, the threatening bellows of a large, successful male were recorded. When the recorded bellows were played to a solitary

Two sea elephants square off in ritual combat. Before these animals fight, they try to frighten off rivals with loud vocalizations.

resting bull, he arose with a start and looked suspiciously about for the source of the sounds. Seeing no challenger, he made no effort to defend himself. But when two bulls were exposed to the recorded threats, each apparently thought the other was responsible and they rose up and began to fight.

111

The Sounds of Predators

Because of their size and power gray whales go unchallenged most of the time. But as man is man's worst enemy, some other whales are the gray whale's enemy. The sleek, substantially smaller orca (the killer whale) is one of these. Tales of orcas attacking whales are generally inaccurate. Healthy adult whales have nothing to fear, but they become very vulnerable when they are sick or when they give birth to their young. The babies themselves are a possible prey during their first year. Orcas emit a great variety of echolocation and communication sounds they use for navigating, keeping contact with the pack, or locating their prey, much as bats do on land. But their whistles and clicks also betray the orcas' presence.

In an experiment conducted off the California coast during the annual migration, the

A gray whale (above) lifts its head out of the water in Scammon's Lagoon in Baja California, a favorite breeding spot for these creatures.

The orca (right) is a cousin of the dolphins. Warm-blooded animals are included in its diet.

recorded sounds of a pack of killer whales were played underwater to a group of approaching grays. Almost immediately some of the surfacing whales wheeled and sped away from the source of the sound. Others, close to forests of kelp, hurried toward the heavy plant growth and hid in it. Such panic can be explained by the fact that the experiment was conducted during the mating and baby-rearing season. In other circumstances, whales and orcas have been seen swimming peacefully together. However, the test clearly establishes that the gray whales can hear, and identify, the calls of the predatory orcas and can stay clear of them.

Chapter VIII. Useful Echoes

To understand how sound can be utilized by man and animals, we must know that sound waves can be bent and reflected. If we remember the alternating compression and expansion of sound traveling through any material, air, or water, we are ready for a new concept—wavelength. If we measure the distance from one compression to the next, this distance is called "wavelength." The higher the frequency, the shorter the wavelength. Since the speed with which sound travels in various materials is known, by knowing the frequency of the sound, we can compute its wavelength.

Much of the sound will be reflected by an object that is large in comparison with the wavelength of that sound. These reflections are called echoes. Sounds are bent around objects which are small compared with a wavelength. The fact that objects will reflect sound has been used by man in more efficiently sounding ocean bottoms.

In 1912 an electrical sound source was built which greatly increased the range of man's underwater sound signaling ability. More electronic devices were developed using the piezoelectric principle—that is, when certain crystals (barium titinate, for example) are bent, they produce an electric current that transforms (or transduces) pressure waves into electronic signals. This principle was first applied by the physicist Paul Langevin in 1917 to the use of electronic acoustics in the sea. Modern versions of these early systems using piezoelectric transducers are the ultrasonic Fathometers, sonars, and fish finders that are found on most modern ships today.

One type of underwater system produces extremely narrow beams of ultrasonic sound, which can be directed like the beam of a searchlight. By scanning the ocean with such devices, we receive a return echo from objects that are large compared to the wavelength of the sound. This system is able to detect surface ships, submarines, fish, and other objects in the ocean, as well as the bottom. If the ocean did not have a surface, a bottom, a varying temperature, and biological organisms, the level of sound would only be decreased by the spreading and absorption of sound waves. But the temperature of the ocean changes with depth, and there are sometimes layers of water with individual temperature (thermoclines). We know that materials of different densities reflect sound, therefore sound in the ocean bounces off the

> "A sound wave sent through the ocean is spread, absorbed, bent, reflected, and scattered."

surface, the bottom, and those masses of water of different temperature as well as animals and plants. Layers of water with different temperatures also cause those sound waves that are not reflected to be bent. Thus a sound wave sent through the ocean is spread, absorbed, bent, reflected, and scattered. The higher the frequency of the sound, the greater the effect. For this reason most long-range ship echo sounders operate at frequencies below 5000 hertz. To detect small objects at much shorter ranges, sonars operate at frequencies beyond the range of man's hearing, that is, above 20,000 hertz. Echo sounders are a fast and reliable source of information.

The blue hole of Lighthouse Reef. Having crossed treacherous shallow waters, Calypso moors in a blue hole, bottomless according to legend, but 460 feet deep according to echo sounders.

Sonar and Polar Navigation

In these pictures a layer of ice two feet or more thick completely covers the Arctic Ocean. The problem encountered by ships much more frequently than a completely covered sea is floating ice (icebergs). When most liquids solidify, their density increases. Water is different though, since when it freezes, it becomes less dense. But the density of ice is only slightly less than water. So most of the iceberg lies underwater, hidden from sight. The problem of sighting ice is compounded by unusual lighting near the poles. The sun is below the horizon during the six months of winter, and during summer the sun remains very low in the sky. This lighting phenomenon can cause visual illusions we call mirages.

To overcome the difficulty of seeing the one-fourth to one-eighth of an iceberg that shows over the surface of the ocean, ships use a variation of the echo sounder called sonar, which is an acronym for SOund Navigation And Ranging. But even with sonar icebergs can only be detected within a distance of a few miles. A few miles may seem like a great distance, but modern, fast-moving ships are difficult to stop or turn, especially large vessels like modern oil tankers.

Now consider the special problem of a polar submariner. His eyes are useless in his dark underwater world. As he is closer to the bottom the depth of the water is even more important to him than it would be to surface ships. To navigate his ship safely, the submariner must rely entirely on the information his electronic equipment gives him about the sea floor below him, underwater cliffs and seamounts in front of him, and the ice above him. Even these obstacles would not be too great if the sea bottom and overhead ice were smooth. His problem is equivalent to that of a pilot of an airplane flying over mountainous country at low altitude in dense fog, except that the aircraft does not have a solid ceiling above it.

Such problems can make passage beneath the polar ice cap seem like traveling through a maze, since at any time a mountain peak or a wall of ice may block a submarine's passage. Without special sonars designed to scan above the submarine as well as ahead and to the sides and down, together with computers that readily process all the data, travel beneath the polar ice caps would be impossible instead of just hazardous.

*Above, **crewmen of** Whale *stretch their legs on the thick layer of ice surrounding their submarine.*

*United States Navy submarine **U.S.S.** Whale (right) surfaces at the North Pole. In a three-dimensional world covered with ice, safety depends upon information from sonar echoes.*

117

Locating Fish by Sound

On the left a ship emits short pulses of high-frequency sound that reflect off the bottom or off objects suspended in the water column beneath the ship. The returning echoes, when graphically recorded, report the actual depth of the bottom and the location of schools of fish. This information can be of great importance to fishermen and scientists. The echogram above is a continuous record of sound reflected off the bottom and off schools of fish as the ship travels along the California coast. Actually a cone-shaped beam of sound emitted from an underwater sound projector (transducer) was directed downward, and the echogram was produced gradually as the ship moved. Sonar uses the same basic principle of reflected sound but offers more possibilities, since the transducer can be moved to project the sound beam in all directions. With this equipment fishermen as well as scientists can not only locate the position and depth of fish schools but estimate their description and size.

Depending on the object of study, different frequencies of sound are used: high frequencies to locate small targets, low frequencies to locate larger targets at greater ranges. Sophisticated large trawlers carry one or more sonar units and a number of echo sounders using a wide range of frequencies to increase their fishing capabilities.

The acoustic deep scattering layer (DSL) is an interesting phenomenon of the sea originally defined by the use of echo sounders. This layer is probably caused by aggregations of marine organisms that reflect sound. At times these layers are so distinct that they had been formerly identified as the ocean bottom. Subsequent to its identification scientists discovered that the DSL makes a daily vertical migration, rising toward the

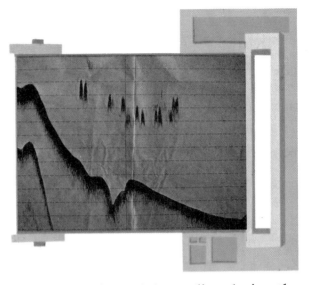

surface at night and descending during the day. Also the DSL may actually consist of a number of distinct layers. Definition of the complex composition of these layers is extremely difficult. It is known that certain fish such as lanternfish form some of these layers but not all of them. In considering what types of marine organisms may form the DSL, we can say what attributes they would have. DSLs have been found in all the oceans of the world, therefore one might suspect that some of the animals involved are worldwide in their distribution. They must be able not only to live at specific depths but be capable of up-and-down movements in response to changes in light level. Finally, they must be able to reflect sound. It is known that in fish this ability is enhanced by the presence of a gas-filled swim bladder found in some species. However, not all fish that are found in the DSL have swim bladders. Another candidate is a group of animals called siphonophores, relatives of the Portuguese man-of-war, which are capable of vertical movement and do have a gas-filled float. The complete composite of the acoustic DSL still presents intriguing mysteries for the ocean scientists who are slowly learning more with the development of more advanced sonars and shipboard computers.

119

Living Sonar

For almost every invention of man an equivalent system exists in nature that far exceeds man's in efficiency and capabilities. One of these is animal sonar, or echolocation.

In 1938 it was discovered that bats emit high-pitched, inaudible sounds, often called ultrasounds (about 40,000-80,000 hertz) and receive echoes that tell them a great deal about their surroundings. Because of the high frequency of its signals, the bat's range of information is moderate but very sharp and accurate. About ten years later, observations of a scientist in Florida led to the discovery of echolocation in dolphins. During an attempt to capture dolphins for a seaquarium, the scientist noted that dolphins could be herded into a canal and in the direction of a net. However, within 100 feet of the unseen net, they abruptly changed direction and swam away. But the dolphins swam into captivity if nets with a larger mesh—or water-soaked ones that had no clinging, sound-reflecting air bubbles—were used.

It was then discovered that dolphins have exceptional abilities to produce and receive sounds with frequencies ranging from a few thousand hertz to over 100,000 hertz. The existence of a dolphin echolocation system was confirmed in the late 1950s; it was then shown that dolphins were able to avoid obstacles, detect and discriminate between small objects, and locate passages in muddy water even on a very dark night. Dolphins were even tested while wearing suction-cup blindfolds.

Dolphins produce two basic kinds of sounds—one is a whistle, which they probably use as a form of communication, and the other is a click, or series of clicks. In addition they can also groan, but nobody knows exactly why. The clicks are thought to be the essential part of the dolphin's sonar. Dolphins have been recorded producing clicks at a rate over 500 hertz. After these sounds have been bounced back by the object they are reflected from, they are received by the dolphins as echoes. The difference between the reflected sound and the emitted sound and the length of time it takes for the echo to return enable the dolphin to learn about its environment. It is possible that the dolphin hears and gains information either by generalizing from a series of clicks or by listening to each individual echo.

There is, of course, no way of knowing what a dolphin hears, but it may be that subtle variations in reflected sound give the dolphin a feeling of what is ahead. The dolphin's system is amazingly accurate and gives the animal many more times the information than is obtained by man from his sonar. For example, navy-trained "Dolly" is capable of retrieving three pennies thrown simultaneously in three different directions; the first is picked up while it is still sinking in midwater, the second and third are found in the sediment, in a few seconds, with a few feet of visibility.

A herd of dolphins frolics across a calm sea. Their built-in sonars keep them constantly aware of their surroundings.

Seeing by Hearing

To observe the dolphin in its own environment, the team of the *Calypso* captured a male while it was riding the bow wave of the ship. Instead of removing it from its natural surroundings, a net "pool" was put in place around the dolphin in the open sea.

He was eventually trained to accept suction-cup blindfolds to test his ability to echolocate without using his visual sense. While blinded, he had no difficulty in navigating within his enclosure. Then the net was replaced by steel rods forming a sort of open cage. The dolphin appeared to search for an opening, but even though he could have escaped through the rods, he remained within the enclosure. The dolphin often emitted a series of clicks while swinging his head from side to side. This enables the dolphin to scan the target with a narrow "beam" of sound, possibly as we would search in the dark with a small flashlight.

Another interesting experiment has shown the high degree of resolution of a dolphin's sonar. A blindfolded dolphin could differentiate a sphere two and one-eighth inches in diameter from a sphere two and one-half inches in diameter. As the test spheres approached each other in size, the correct responses of the dolphin decreased. A two-and-one-quarter-inch sphere could be identified as different from a two-and-one-half-inch sphere 70 percent of the time. However, beyond a certain point the dolphin refused to continue the experiment; perhaps he knew the limitations of his system and refused to guess.

Placing suction cups. A diver blindfolds a dolphin (left) to test its ability to navigate by sonar.

Taking off. The blindfolded dolphin (below) is ready to explore the area using pulses of sound.

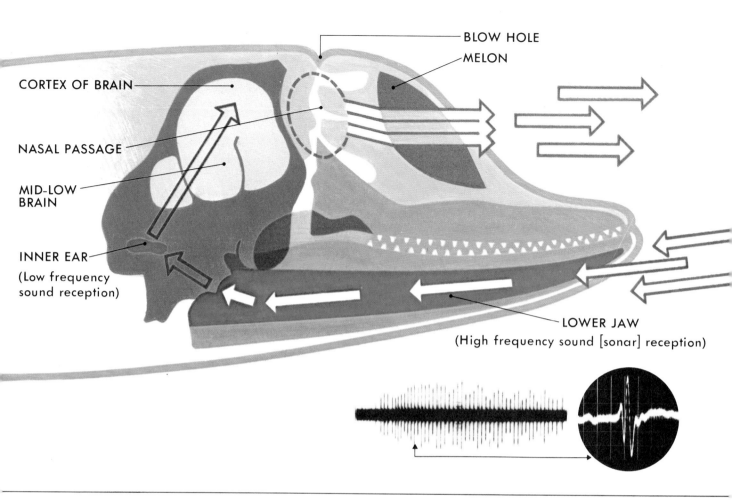

Dolphin's Sonar

To obtain information about their environment, dolphins emit clicking sounds of frequencies ranging from less than 2000 to more than 100,000 hertz. These clicks may be given as individual sounds or as trains of sounds strung together. The dolphin, and other members of the order of toothed whales, can determine not only the range and bearing, but also the size, shape, texture, and density of objects. It may also be able to perceive more information than we can simply by varying the pitch of individual clicks of each train, and each echoed click, being different, may bring back a different message. Thus one single train of echoes gives a composite mental image of an object. It is the character of those clicks as modified into an echo by the target that informs the dolphin of the object's makeup.

There are at least four types of information in the echo: the direction from which the echo returns, the change of frequency, the amplitude of the sound, and the time elapsed from emission to return. As the dolphin scans with its head acting as a biological sonar transducer, it can determine the direction from which the echoes are returning and thus the bearing of the object under scrutiny. The changes of frequency tell about its size and shape. The sound's amplitude and the time elapsed give clues to the distance.

If the dolphin is one of a school all of whom may be echolocating simultaneously, the noise level may be extraordinarily high. However, the dolphin's ability for directional hearing plus its ability to filter out extraneous sounds (which it shares with humans) probably make it possible to perceive its own clicks and echoes above all the other sounds.

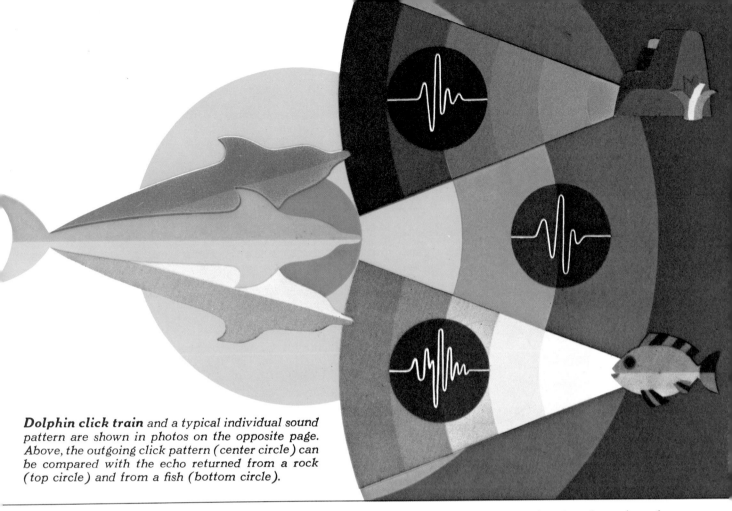

Dolphin click train and a typical individual sound pattern are shown in photos on the opposite page. Above, the outgoing click pattern (center circle) can be compared with the echo returned from a rock (top circle) and from a fish (bottom circle).

How the clicks are produced and emitted and how the dolphin perceives the echoes is only now beginning to be understood: the clicking emissions originate inside the dolphin's head. The sounds are produced even while the animal is underwater without loss of air. suggesting that the air is recycled within the dolphin's respiratory tract.

The sides of a dolphin's head and its lower jaw, containing an oily fat, are the areas that receive echoes. Hearing is so precise that the animal can discern differences between objects of similar size, shape, and texture and can even recognize different metals.

When a dolphin is traveling, it usually moves its head slowly from side to side and up and down. This motion is a sort of general scanning; it enables the dolphin to "see" a broader path ahead of it. But then if it gets interested in a small target, such as a fish in murky water, its scanning head motions become fast and jerky. The explanation finds roots in the facts that low frequencies are far-reaching but not directional, and high pitch clicks are for short-range, high-definition investigations.

Unlike high-frequency sound, low-frequency vibrations are probably received initially in the inner ear. To receive and interpret all these echoes, the dolphin's brain has a much larger auditory lobe than our brain.

Among the mysteries yet to be unraveled are how the dolphin deals with those marine acoustical problems that so drastically limit the performances of man-made sonars: refraction, reverberation, scattering, and reflection of sound waves underwater, and the effects of various water characteristics, such as density, temperature, and the presence of plankton in large concentrations.

Belugas: Canaries of the Sea

The beluga is an arctic whale whose color matches the snow and ice of its habitat. Beluga in Russian means white. It is closely related to the small whale known as the narwhal, which, because of an enormously long upper incisor, is called the unicorn of the sea. Belugas grow to a length of about 15 feet and may live to be close to 50 years old. When they are born they are blue-gray in color. Their color fades until by the age of 4 or 5 they are pale yellowish white. A few years more and they turn a pure white. Belugas' forms are whalelike, but they lack a dorsal fin.

Probably the most outstanding physical feature of the beluga whale is its lopsided, bulbous head. The asymmetry of the creature's head is readily noticeable because the blowhole is located just to the left of the center line of the head. The bulbous shape of the forehead on beluga whales resembles the "melon" on the dolphin's head. It is the same in structure and probably the

through turbid waters. The beluga's ability to echolocate is probably equal to that of the dolphin, and far superior to man's most sophisticated echo sounders.

The beluga is such a vocal creature that its underwater phonations can be heard above the surface of the sea. Besides the trains of clicks, belugas also produce pure tone whistles usually associated in other cetaceans with communication. Some of their underwater phonations are described as sounding like chirping. It is the chirping and the pale yellowish color of the juvenile whales that have won them the name of sea canaries. Some of their other phonations, however, have been described as growls, screams, squeals, and roars. Belugas have been seen swimming just below the surface, screaming, squealing and emitting a stream of bubbles from their blowholes. These phonations, probably produced by constricting air passages that lead to the blowhole and exhaling, have been measured through a wide range of sound from about 500 hertz to close to 10,000 hertz.

*A pair of **beluga whales** in captivity (left). Few aquariums succeed in keeping belugas in good health.*

***Bulbous forehead.** The muscular "melon" on the forehead of the beluga (below) bulges as the animal emits clicking sounds.*

same in function as the dolphin's melon. The melon on both animals is very likely the source of echolocation clicks. Belugas in captivity have been observed while emitting clicking noises to bulge out their forehead muscles and to indent those just below. The complex mass of muscles in the beluga's forehead may enable it to beam sounds in the direction of its heading. Although there is no firm proof yet that belugas use echolocation, it seems apparent that they do. They are able to navigate over long distances

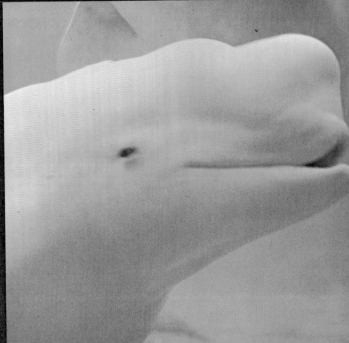

Standing on its tail. *A spy-hopping gray whale stands on its tail in a shallow lagoon in Baja California. Orcas and sperm whales also spy hop.*

Beached. *A huge right whale is found stranded on the shore. The animal may have been ill or its echo navigation system may have failed.*

Whale Riddles

Each winter gray whales migrate southward along the California coast, swimming in relatively shallow water. Occasionally a gray does a tail stand—called spy hopping—lifting its huge head high above the surface. We don't know why they do it, but some observers have suggested the whales are getting their bearings, navigating by the shoreline, and deciding what direction to swim in. If this is so, if they need to look around to find their way, then these whales must be equipped with a poor natural sonar, if any. In fact, gray whales do not emit the clicks we believe are associated with echolocation.

Even animals that do use sonar falter at times and lose their lives as a result. Lone whales are sometimes stranded in shallow water when they become ill, or when their echo navigation system somehow fails to give them proper directional information. When it is ill, a whale may panic, and swim even further into the shallow water trap.

Many whales are social animals and travel in groups, or pods. If the leader of the pod finds itself in shallow water, its companions probably follow it toward the shore until all the members of the group are beached. This probably accounts for the multiple strandings of apparently healthy animals. Sea floor characteristics may confuse them.

Chapter IX. Is There Another Language?

Language is a communication of thoughts and feelings. Man is unique in the animal kingdom in being able to communicate through specific, well-defined vocal patterns as well as through written transcriptions of his expressions. Less-developed communication patterns exist in nearly all forms of life, transmitted by chemical stimuli, touch, smell, taste, and so on. Identifying these forms of communication has always been a

> "The question is: Are there any animals besides man with a language as we define it?"

difficult problem. The question is: Are there any other animals besides man with a language as we define it? Research has found hundreds of different types of sounds in the sea—from clicks and pops to choruses of many animals together. Are these animals actually "talking"? Cetaceans, we know, are highly social animals, and they communicate with one another, but when does this communication begin to be classified as language?

On land there is no animal equipped with a brain comparable to man's. But in the sea there are several mammals, including orcas, sperm whales, dolphins, and porpoises, that have brains which are at least anatomically equal to man's in size. They are the only creatures on earth to be gifted with the nervous system potentially capable of higher thought processes. The same animals happen to have the ability to produce a wide variety of sounds. This is not the case for the dog (small brain, limited voice), the apes (small brain, limited voice), the parrot (voice but small brain), and so on.

We have good reasons for thinking man had a substantial language 30,000 years ago, when his brain was perhaps less developed than that of today's dolphin. For these reasons, and a few others, there are some scientists today who are trying to determine if animals do have a language and talk to each other. Dolphins have the brain and an extensive sound repertoire. We are now trying to find out what they are saying to each other and how well defined their ability to communicate is. Two isolated dolphins that are linked by phone constantly exchange many sounds and rarely overlap each other. Some captive dolphins have reshaped their sounds to mimic the whistles of men—perhaps attempting to establish a basis for interspecies communication. This ability to manipulate sounds is encouraging, but we should not forget that a parrot can also mimic human sounds and produce them on cue. Can dolphins communicate to each other not only that food is available, but more specific messages about the food? Can they express their preferences to one another, and as they belong to the aristocracy of creatures that spend most of their lives playing, can they also develop the kind of abstractions a language is made of? There are many pitfalls in researching this subject. It is all too easy to assume that a particular vocalization means that a complex idea is being communicated rather than an expression of fear, happiness, displeasure, annoyance, sexual arousal, or conflict. It may be that man is, in fact, alone. Further scientific study may give us the answer.

Underwater communication devices. A working diver wears a communication device which lets him talk to the boat crew on the surface as well as to other divers within a range of 400 to 500 feet.

Songs of Humpback Whales

Every spring humpback whales pass through the clear blue waters around Bermuda. The whales are on their way to the rich feeding grounds of the Arctic Ocean. On calm days, when there are no crashing waves to obscure the sounds, fishermen in silent sailboats may be treated to one of nature's most beautiful melodies—the songs of the humpback whales.

Whales are vocal animals. The humpbacks are noisiest of all baleen whales, singing their songs day and night during their long migration to and from warm southern waters. The vocalizations are rightly called songs; they occur in complete sequences, which are repeated again and again. A song may be as short as six or seven minutes or as long as 30 minutes. A humpback may put several songs

A mother and baby **humpback whale** *(above) in Bermuda waters cruise along just beneath the surface. The air bubbles drifting toward the surface are from the diver-photographer, not the whales.*

Courting. *During the migrations males and females may look for mates. Males sing to convey their intentions to fertile females. With their vocalizations the females encourage or discourage suitors.*

together without a break, thereby giving an extended performance which may last for hours.

At first, observers were confused about the humpback's songs, since they didn't seem to follow any pattern. To simplify their task, the observers recorded the songs and deciphered them with a spectograph. Through analysis they discovered that the songs can be broken down into smaller units. These are called themes, and it was discovered that the songs of any whale, although different in

length, always have the same number of phrases in a theme, and a phrase may be repeated any number of times. With each rendition of a single phrase it is changed slightly until, after many repetitions, the final phrase is completely different from the original one.

Males and females both sing the songs. The reasons for the performances are still unclear. Because of the difficulty in observing the whales and recording their songs, we have not yet learned the behavior associated with the singing. But whales respond to the songs of another. We also know that the whales do not sing as much once they arrive at their summer feeding grounds in the polar regions.

The whales migrate as a loose community, spread over miles of open ocean. Perhaps the songs help the whales stay together as they move. Since the sounds travel far, and the whale's hearing is keen, the migrants can keep track of their neighbors and keep from falling behind.

Giving support. A sperm whale attended by its companions after being injured by a propeller.

Organized Sperm Whales

Sperm whales travel in groups, or pods, of 10 to 30—a dominant male is followed by a harem of females and young calves. During one of the early expeditions aboard *Calypso,* the ship encountered a group of sperm whales. One of the whales suddenly veered right into the ship's path. The entire hull shuddered from the impact with the huge

mammal. The force of the collision seemed to cause the whale no major injury. The subdued chirping of the whales before the collision changed suddenly, and the noises became more frantic and excited. All the whales in the group seemed to learn of the collision in an instant, perhaps through the language of their mouse squeaks. Whales arrived on the scene from all directions. Some of them rushed to the side of the stricken whale to

Abandoned. After assessing the damage, the whales abandoned the fatally injured baby.

support it at the surface, keeping its blow-hole out of the water so it could breathe.

The agitated chirping calmed as the stricken whale resumed normal breathing. But then a baby whale drifted into one of *Calypso's* spinning propellers. Five great gashes were carved into its flesh. It returned to the group and was surrounded and supported as the first wounded whale had been. Suddenly the largest whale lifted its massive head out of the water to glare at the ship. As if on signal, as if a medical diagnosis had been made, the whales dispersed, abandoning the fatally injured baby.

Such disciplined and organized behavior indicates that whales respond to messages they receive, even when they are out of range of vision.

Experiments with dolphins. *A hydrophone suspended in a temporary open-ocean cage allows researchers to listen to dolphins.*

Responding. *A dolphin jumps high in response to a hand signal. Many people believe that dolphins have their own language.*

Experiments with Language

When two or more dolphins are housed in a single tank, they whistle to each other. What the whistles mean to the dolphins is unknown, but it has been observed that the sounds have a noticeable effect on all the dolphins in the tank, even the ones not doing the whistling. If one dolphin whistles, as though initiating a "conversation," other dolphins in the tank respond by whistling back. Dolphins exchange whistles in the wild as well as in captivity.

The sound of a dolphin's exhalation is normally a subdued "whoosh," but in times of stress or excitement the noise can be quite loud. When captive dolphins do this, other dolphins in nearby tanks may also exhale loudly, even though they are not being subjected to the same stress. Their loud breathing seems to be in response to the sounds of the exhalations of the first dolphin.

Some cetacean researchers disregard the formidable complexity of the dolphin's brain and suggest that their level of communication is no higher than that of other advanced mammals or of highly social birds. If this was the case, dolphins would communicate simple desires about food or mating or warnings about the presence of predators, but little more. Other scientists contend that dolphins can actually communicate complex

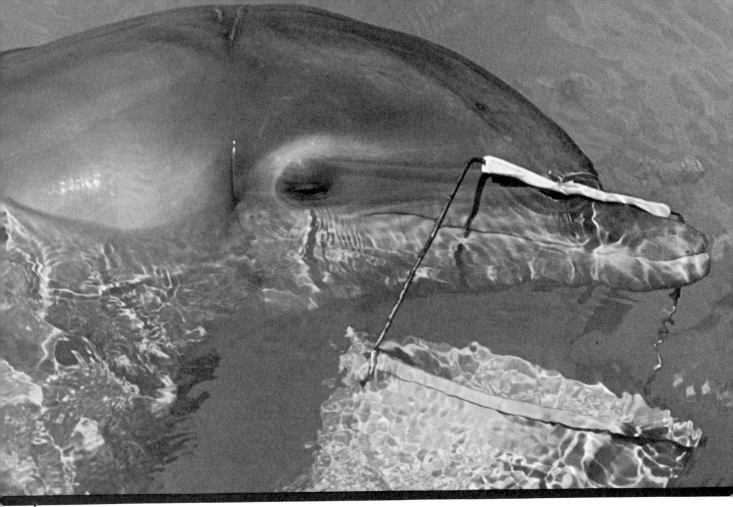

*A **bottle-nosed dolphin** retrieves a pail on a signal from its trainer. Dolphins have proved to be adept, willing pupils.*

thoughts through their vocalizations. Instead of simply telling their companions that food is available, they may also be able to express moods that have little to do with the basic motivations of simple animals, food, survival, or sex. But this is only speculation. Perhaps we are just expecting the dolphins to communicate like humans, since they act remarkably like us in many situations.

Experiments have been conducted in hope of proving that dolphins do communicate and exchange ideas. Two dolphins have been placed in adjacent tanks and a "telephone" consisting of a transmitter and a receiver submerged in each tank. Without the telephone the dolphins could not hear each

other, but with it their vocalizations were transmitted electronically and they could talk back and forth. They could not see each other. The dolphins exchanged clicks and whistles for most of the time the telephone was turned on. When one vocalized, the other remained silent. This pattern seems to indicate that the dolphins were conversing, perhaps communicating, but again the meaning of the whistles is unknown. When the telephone was turned off, the dolphins stopped producing a variety of whistles and emitted only "signature whistles," repeating them over and over. These "signature whistles" are personal whistles believed to enable other dolphins to recognize individual animals.

Sounds of Dolphins

Some scientists have been encouraged by their experiments with dolphins, since these animals seem as anxious to communicate with people as we are to communicate with them. Dolphins are also apparently willing to try to communicate in the medium of our choosing. When human voices are electronically converted into the clicks and whistles dolphins use, and are transmitted underwater, the dolphins answer them underwater. If we speak to them in air, they vocalize in air. This is especially noteworthy since dolphins rarely, if ever, produce sounds in air when they are in the wild. If dolphins have a language, or languages, it may not be a complex one, but merely a collection of simple sounds. That they are communicating and not just making random noises seems likely, since the vocalizations between two animals are not monotonous repetitions of the same sounds. Each is different from the last, but they are emitted in a structured pattern. Interestingly, young dolphins use much simpler sounds than older dolphins. Could it be that they learn more "words" as they grow older?

Dolphins have learned to mimic man's speech, but this is not really communication. Their words are simply parrotlike repetitions of phrases they have heard again and again. Although the speech resembles ours, there is no evidence that the dolphins know what they are saying. Recent work has led many scientists to believe that dolphins are about as intelligent as dogs, and for this reason believe man-dolphin communication will probably never exceed man-dog communication. Some researchers, however, still believe that men and dolphins may eventually be able to communicate on a higher level than this.

Whistle Language

The Canary Islands are the tips of volcanic mountains which rise thousands of feet almost vertically from the floor of the Atlantic Ocean. When hot lava spilled from volcanic vents to form the mountains, an extremely varied terrain was created, with high mountains separated by deep ravines.

Transportation by foot is the primary means of getting from one place to another on the rugged island of Gomera. Scrub brush blanketing much of the island's surface makes travel over the up and down terrain difficult. Should people high up on a mountain's slope want to communicate with someone far below, it is a long and tiring walk down. If they try shouting, their voices are carried away or garbled by the wind. So, to speak with one another without having to scramble over the rugged ground, the people of Gomera Island have developed a complex whistle language.

With whistle language Canary Islanders on different mountains can communicate with one another, or they can "talk" to people in valleys far below.

The islanders whistle through their fingers (inset) manipulating their lips and tongues as when they speak. It is a language in itself, unrelated to Spanish.

The language has evolved over hundreds of years. It is not a code, as once thought, but a true language. The whistles sound something like birds' songs, totally unrelated to Spanish. To whistle, two fingers are inserted into the mouth, while lips and tongue modulate the trebles. Very similar whistle languages have been developed independently by shepherds in the Pyrennees and in Turkey. Our languages are all *articulated,* in contrast to the whistles which are *modulated,* and these may be the keys to comprehensive computer programming that might ultimately lead to understanding dolphins.

139

Trapping Porpoises and Tuna

There is no longer much doubt in anyone's mind that porpoises are able to communicate. Yet for all their ability to "talk" to each other, at least on the simplest terms, porpoises die by the hundreds of thousands each year in the nets of commercial tuna fishermen off Central and South America.

The porpoises die when they are caught in the fishing nets put out to catch yellowfin and skipjack tuna. The tuna live in close association with porpoises, and fishermen know it. They leave port to catch tuna, equipped with a fishing vessel known as a purse seiner of about 250 feet, two to four ponga boats, a net skiff, and an enormous fishing net called a purse seine, which can be drawn together along one edge with a rope line that acts like a drawstring on a purse.

The fishing captain serves as lookout in the seiner's crow's nest. He spots the porpoises, ascertains the presence of the tuna with them, and directs the entire operation by radio. The ponga boats are sleek, 16-foot speed boats, which race around the porpoise herd to round them up into a compact group, like cowboys in a cattle roundup. The tuna stay close beneath the porpoises, as they are herded together. Then the net skiff, a powerful 25-footer, is launched from the stern of the purse seiner, which serves as mother ship for this team. The purse seiner carries one end of the mile-long seine and circles the porpoises and tuna, while the ponga boats keep the herd from splitting into smaller groups.

If the porpoises were as intelligent as they seemed to be, and if they did communicate with each other, why couldn't they warn each other of the approaching danger of the net? Some of the porpoises do flee before the seine has encircled them. Why is it that those able to escape do not "tell" the other porpoises how to? Later, when the fishermen are mak-

ing an effort to free the porpoises before hauling in the tuna net, some additional porpoises escape. Why don't they tell the others how to get out? Do they panic like people in a flaming building stampeding to escape? To separate the many porpoises from the tuna, many experiments are being carried out. In one the recorded calls of orcas are played underwater during fishing operations. The hope is that the recorded sounds will frighten the porpoises out of the nets because orcas include dolphins and porpoises in the long list of their menus, and are feared by them. Dr. James Fish of the Bioacoustical Branch of the Naval Undersea Center in San Diego feels that his experiments with playback of orca sounds may help reduce the number of porpoises that die in the tuna nets. His optimism is guarded, however, because experiments are still inconclusive. To start with, how do we know for sure that the orca calls that are selected for playback are

really the aggressive ones. Many more trials are to be undertaken.

Many land mammals and a number of birds have a handful of calls, songs, or cries that indicate basic needs or signal simple messages —food and feeding calls, courtship and mating songs, threats to territorial invaders and alarm calls. Porpoises too sound the alarm to each other, but it is puzzling that when encircled by the tuna nets they fail to escape without man's help. The sound of the ponga boats racing around above them may terrorize them to such a degree that they become unable to sound the alarm or to help their companions out of the trap. Yet a female porpoise occasionally has been known to re-enter the net in search of its calf. The loud buzzing of the ponga boat propellers certainly confuses the entire acoustic system of the porpoises. The *Calypso* divers have often used the technique of such fast and noisy circling to neutralize a whale as large as a sperm whale. The animal slows down, stops diving and remains acoustically "hypnotized" as long as the merry-go-round lasts.

Another tragedy happens on a much smaller scale off Provincetown, Massachusetts, where commercial fishermen set large fish traps. In the waters of that area small schools of the little harbor porpoise are not uncommon. Occasionally one individual from a school swims into one of the fishermen's traps and is unable to escape. Again why is it that the companions of the trapped porpoise do not tell him how to get out?

Researchers continue their efforts to find answers to these and other questions about porpoises, dolphins, and whales.

*A **purse seining tuna vessel** heads out to the fishing waters off Central America. Following it are ponga boats which aid in the roundup of the fish.*

Vibrating with the Sea

Today I know I am invited in the sea.
Far in the Atlantic, dolphins, for an instant
Suspend their runs to stare at me and flaunt their skills.
I step in tenderly skin to skin with the sea.
In crystal-clear water the range of my sight shrinks
To the size of a church all built in stained glass blue . . .

Silence-minded I breathe as softly as a purl
I glide down—further down—a threatening night settles.
Ultramarine glimmers filter in from nowhere.
A start of consciousness—I arrest my descent—
From heroic recalls of forgotten reasons
Safety—computations—all from another World.
No more able to tell Up from Down—North from South
Surface has long vanished, dolphins have fled away.

A sudden touch of fear—in trance I close my eyes
Behind my ears I feel the pull of gravity
That water filched from me by way of wedding gift.
I stop breathing to hear my dolphin's distant songs
I half open my lips and salt wakes up my taste.
I prod a drifting salp to sense its flimsy flesh.

When I raise my eyelids, under me, around me,
In closing circles: flights of tuna, lonesome sharks . . .
I could not smell their musk and my skin has not sensed
The mighty swimming strokes that beat like jungle drums.

Raving I stretch my arms to embrace all the sea
My rapture transcends me in foolish ecstasy
My chest now resounding from remote restless tails
Soon I would obey the call of a migration
And all sensors aroused I would rejoin my kind—
Our strong and lissome flanks would whip as One with pride
Vibrating with the sea, all in tune with the world.

Index

ILLUSTRATIONS AND CHARTS:

Howard Koslow—13, 14-15, 39, 42-43, 124; Living Design Corporation, Walter Monczka—93, 118-119, 125.

PHOTO CREDITS:

Ken Balcomb—110-111, 126-127 (top); John Boland—104; Pat Colin—90-91; David Doubilet—26 (right), 61, 69, 84; Jack Drafahl, Brooks Institute of Photography—23, 24, 31, 64, 85; James Dugan—11, 134, 135; William Evans, Naval Undersea Center, Washington, D.C.—120-121; Freelance Photographers Guild: B. W. Brown—66, James Dutcher—57, 63, FPG—107, Bob Gladden—105, Dennis Hallinan—138, J. W. LaTourrette—36, Ben McCall—127 (bottom), Tom Myers—2-3, 28, 38, 76, 92, Chuck Nicklin—27, 35, John Stormont—32, Western Marine Laboratory—65; Allan Gornick—131; Al Grotell—68; International Imaging Systems—47; Steve Leatherwood—112, 128 (left), 140-141; George Lingle—142; Kenneth F. Mais, California Department of Fish and Game—119; Maltini-Solaini, M. Grimoldi, Rome—56, 70-71, 83, 106; Marineland of Florida, St. Augustine, Fla.—96-97; Richard Murphy—17 (right), 37, 62, 78, 79, 99; NASA—50-51; Naval Photographic Center, Washington, D.C.—34, 116, 117; Naval Undersea Center, San Diego, Calif.—52-53, C. Scott Johnson, 87; Donald R. Nelson—100-101; Chuck Nicklin—72-73, 128-129 (right); Carl Roessler—22, 30, 55, 59, 60; Sea Library: Carl Roessler—83; C. L. Smith—102-103; Tom Stack & Associates: R. H. Burrell—113, Ron Church—108-109, Ben Cropp—25, 29, E. R. Degginger—67, Warren Garst—94-95, Keith Gillett—21, 80, 81, Ben Goldstein—5, Bill Noel Kleeman—12, Jack McKenney—26 (left); William M. Stephens—58; Valerie Taylor—75; Ed Zimbelman—19.